ORTHOPEDIC CLINICS OF NORTH AMERICA

Sexual Dimorphism in Musculoskeletal Health

GUEST EDITORS
Laura L. Tosi, MD
Letha Y. Griffin, MD, PhD
Mary I. O'Connor, MD

October 2006 • Volume 37 • Number 4

An Imprint of Elsevier, Inc.
PHILADELPHIA   LONDON   TORONTO   MONTREAL   SYDNEY   TOKYO

W.B. SAUNDERS COMPANY
A Division of Elsevier Inc.

Elsevier Inc., 1600 John F. Kennedy Blvd., Suite 1800, Philadelphia, PA 19103-2899.

http://www.orthopedic.theclinics.com

ORTHOPEDIC CLINICS OF NORTH AMERICA
October 2006
Editor: Debora Dellapena

Volume 37, Number 4
ISSN 0030-5898
ISBN 1-4160-4763-8

Copyright © 2006 by Elsevier Inc. All rights reserved. No part of this publication may be reproduced or transmitted in any form or by any means, electronic or mechanical, including photocopy, recording, or any information retrieval system, without written permission from the Publisher.

Single photocopies of single articles may be made for personal use as allowed by national copyright laws. Permission of the Publisher and payment of a fee is required for all other photocopying, including multiple or systematic copying, copying for advertising or promotional purposes, resale, and all forms of document delivery. Special rates are available for educational institutions that wish to make photocopies for non-profit educational classroom use. Permissions may be sought directly from Elsevier's Rights Department in Philadelphia, PA, USA: phone (+1) 215 239 3804, fax (+1) 215 239 3805, e-mail healthpermissions@elsevier.com. Requests may also be completed on-line via the Elsevier homepage (http://www.elsevier.com/locate/permissions). In the USA, users may clear permissions and make payments through the Copyright Clearance Center, Inc., 222 Rosewood Drive, Danvers, MA 01923, USA; phone: (978) 750-8400, fax: (978) 750-4744, and in the UK through the copyright Licensing Agency Rapid Clearance Service (CLARCS), 90 Tottenham Court Road, London W1P 0LP, UK; phone (+44) 171 436 5931; fax: (+44) 171 436 3986. Other countries may have a local reprographic rights agency for payments.

The ideas and opinions expressed in *Orthopedic Clinics of North America* do not necessarily reflect those of the Publisher. The Publisher does not assume any responsibility for any injury and/or damage to persons or property arising out of or related to any use of the material contained in this periodical. The reader is advised to check the appropriate medical literature and the product information currently provided by the manufacturer of each drug to be administered to verify the dosage, the method and duration of administration, or contraindications. It is the responsibility of the treating physician or other health care professional, relying on independent experience and knowledge of the patient, to determine drug dosages and the best treatment for the patient. Mention of any product in this issue should not be construed as endorsement by the contributors, editors, or the Publisher of the product or manufacturers' claims.

Orthopedic Clinics of North America (ISSN 0030-5898) is published quarterly (For Post Office use only: Volume 37 issue 4 of 4) by Elsevier Inc., 360 Park Avenue South, New York, NY 10010-1710. Months of publication are January, April, July, and October. Business and Editorial Offices: 1600 John F. Kennedy Blvd., Suite 1800, Philadelphia, PA 19103-2899. Customer Service Office: 6277 Sea Harbor Drive, Orlando, FL 33887-4800. Periodicals postage paid at New York, NY and additional mailing offices. Subscription prices are $205.00 per year for (US individuals), $347.00 per year for (US institutions), $243.00 per year (Canadian individuals), $407.00 per year (Canadian institutions), $281.00 per year (international individuals), $407.00 per year (international institutions), $103.00 per year (US students), $140.00 per year (Canadian and international students). Foreign air speed delivery is included in all *Clinics* subscription prices. All prices are subject to change without notice. **POSTMASTER:** Send address changes to *Orthopedic Clinics of North America*, Elsevier Periodicals Customer Service, 6277 Sea Harbor Drive, Orlando, FL 32887-4800. **Customer Service: 1-800-654-2452 (US). From outside of the US, call 1-407-345-4000. E-mail: elspcs@elsevier.com.**

Reprints. For copies of 100 or more, of articles in this publication, please contact the Commercial Reprints Department, Elsevier Inc., 360 Park Avenue South, New York, New York 10010-1710. Tel. (212) 633-3813 Fax: (212) 462-1935 e-mail: reprints@elsevier.com.

Orthopedic Clinics of North America is covered in *Index Medicus, Cinahl, Excerpta Medica, and Cumulative Index to Nursing and Allied Health Literature.*

Printed in the United States of America.

SEXUAL DIMORPHISM IN MUSCULOSKELETAL HEALTH

GUEST EDITORS

LAURA L. TOSI, MD, Associate Professor of Orthopaedics and Pediatrics, George Washington University; Director, Bone Health Program, Division of Orthopaedics, Children's National Medical Center, Washington, DC

LETHA Y. GRIFFIN, MD, PhD, Peachtree Orthopedic Clinic, Atlanta, Georgia

MARY I. O'CONNOR, MD, Associate Professor and Chair, Department of Orthopedic Surgery, Mayo Clinic, Jacksonville, Florida; Chair, American Academy of Orthopaedic Surgery Women's Health Issues Advisory Board

CONTRIBUTORS

GUNNAR B. J. ANDERSSON, MD, PhD, Professor and Chairman of Orthopaedic Surgery, Department of Orthopaedic Surgery, Rush University Medical Center, Chicago, Illinois

ELIZABETH A. ARENDT, MD, Professor and Vice-Chair, Department of Orthopaedic Surgery, University of Minnesota, Minneapolis, Minnesota

JUDITH F. BAUMHAUER, MD, Professor of Orthopaedics and Chief, Division of Foot and Ankle Surgery, University of Rochester School of Medicine, Rochester, New York

BRUCE BEYNNON, PhD, Department of Orthopaedics and Rehabilitation, University of Vermont College of Medicine, Burlington, Vermont

ADELE L. BOSKEY, PhD, Professor, Biochemistry; Professor, Physiology, Biophisics, and Systems Biology, Weill Medical College of Cornell University; Starr Chair in Mineralized Tissue Research, Hospital for Special Surgery, New York, New York

BARBARA D. BOYAN, PhD, Price Gilbert, Jr. Chair in Tissue Engineering, Wallace H. Coulter Department of Biomedical Engineering, Georgia Tech and Emory University, Georgia Institute of Technology, Institute of Bioengineering and Bioscience, Atlanta, Georgia

GWYNNE BRAGDON, MD, Orthopaedic Resident, University of Rochester School of Medicine, Rochester, New York

JULIA CHAN, BA, Doctoral Student, Division of Gerontology, Department of Epidemiology and Preventive Medicine, University of Maryland School of Medicine, Baltimore, Maryland

DAVID FEINGOLD, MD, Clinical Instructor, Division of Sports Medicine, Department of Orthopaedic Surgery, The David Geffen University of California Los Angeles School of Medicine, Los Angeles, California

EDWARD J. GOLDBERG, MD, Assistant Professor of Orthopaedic Surgery, Department of Orthopaedic Surgery, Rush University Medical Center, Chicago, Illinois

SHARON L. HAME, MD, Assistant Professor, Division of Sports Medicine, Department of Orthopaedic Surgery, The David Geffen University of California Los Angeles School of Medicine; The Greater Los Angeles Veteran's Administration Hospital, Los Angeles, California

JO A. HANNAFIN, MD, PhD, Orthopaedic Director, Women's Sports Medicine Center, Director of Orthopedic Research, Hospital for Special Surgery; Professor of Orthopedic Surgery, Weill Medical College of Cornell University, New York, New York

DANIEL M. HARDY, PhD, Department of Cell Biology and Biochemistry, Department of Orthopaedic Surgery and Rehabilitation, Texas Tech University Health Sciences Center School of Medicine, Lubbock, Texas

JOHN R. HICKOX, MS, Department of Cell Biology and Biochemistry, Department of Orthopaedic Surgery and Rehabilitation, Texas Tech University Health Sciences Center School of Medicine, Lubbock, Texas

MARY ANN KEENAN, MD, Chief, Neuro-Orthopaedics Program, Professor and Vice Chair for Graduate Medical Education, Department of Orthopaedic Surgery, The University of Pennsylvania, Philadelphia, Pennsylvania

JOSEPH MICHAEL LANE, MD, Department of Orthopedic Surgery, Weill Medical College of Cornell University; Chief, Metabolic Bone Disease Service, Hospital for Special Surgery, New York, New York

MICHAEL D. LOCKSHIN, MD, Director, Barbara Volcker Center for Women and Rheumatic Disease; Co-Director, Mary Kirkland Center for Lupus Research; Professor of Medicine and Obstetrics-Gynecology, Joan and Sanford Weill Medical College of Cornell University; Hospital for Special Surgery, New York, New York

JAY MAGAZINER, PhD, MSHyg, Professor and Director, Division of Gerontology, Department of Epidemiology and Preventive Medicine, University of Maryland School of Medicine, Baltimore, Maryland

ANNA-LENA MAKOWSKI, HTL, Research Associate, Department of Orthopaedics, University of Miami Miller School of Medicine, Miami, Florida

NEIL A. MANSON, MD, FRCSC, Orthopaedic Spine Fellow, Department of Orthopaedic Surgery, Rush University Medical Center, Chicago, Illinois

MARY I. O'CONNOR, MD, Associate Professor and Chair, Department of Orthopedic Surgery, Mayo Clinic, Jacksonville, Florida; Chair, American Academy of Orthopaedic Surgery Women's Health Issues Advisory Board

KATHRYN O'CONNOR, PT, Medical Student, University of Rochester School of Medicine, Rochester, New York

DENISE L. ORWIG, PhD, Assistant Professor, Division of Gerontology, Department of Epidemiology and Preventive Medicine, University of Maryland School of Medicine, Baltimore, Maryland

E. ANNE OUELLETTE, MD, MBA, Chief of Hand Surgery and Professor, Department of Orthopaedics, University of Miami Miller School of Medicine, Miami, Florida

VIVIAN W. PINN, MD, Director, Office of Research on Women's Health, National Institutes of Health, Department of Health and Human Services, Bethesda, Maryland

CATHLEEN L. RAGGIO, MD, Assistant Attending, Hospital for Special Surgery, New York, New York

BRADLEY RAPHAEL, MD, Intern, Department of Orthopedics, Hospital for Special Surgery, New York, New York

ALANA CAREY SEROTA, MD, CCFP, Fellow, Department of Orthopedics, Hospital for Special Surgery, New York, New York

MONIQUE A. SHERIDAN, BA, Student, University of Maryland Medical School, Baltimore, Maryland; Formerly, Ludwig Research Fellow, Women's Sports Medicine Center, Hospital for Special Surgery, New York, New York

JAMES R. SLAUTERBECK, MD, Department of Orthopaedics and Rehabilitation, University of Vermont College of Medicine, Burlington, Vermont

KIMBERLY J. TEMPLETON, MD, McCann Professor of Women and Science, University of Kansas Medical Center, Kansas City, Kansas

LAURA L. TOSI, MD, Associate Professor of Orthopaedics and Pediatrics, George Washington University; Director, Bone Health Program, Division of Orthopaedics, Children's National Medical Center, Washington, DC

CONTENTS

Preface xi
Laura L. Tosi, Letha Y. Griffin, and Mary I. O'Connor

Past and Future: Sex and Gender in Health Research, the Aging Experience, and Implications for Musculoskeletal Health 513
Vivian W. Pinn

> The statistics about common musculoskeletal disorders describe a few of the many health conditions that affect men and women. For such disorders and conditions, there are differences in incidence, predisposition, and therapeutic and preventive strategies for managing them. Although we have made progress in women's health research, many challenges remain, including those related to conditions and diseases of the musculoskeletal system that may affect women and men differentially. Research is needed to identify genetic, hormonal, environmental, and societal factors that contribute to these sex and gender differences and to understand when appropriate clinical applications should differ or be the same.

Does Sex Matter in Musculoskeletal Health? A Workshop Report 523
Laura L. Tosi, Barbara D. Boyan, and Adele L. Boskey

> In April 2004, the American Academy of Orthopaedic Surgeons, the National Institute of Arthritis and Musculoskeletal and Skin Diseases of the National Institutes of Health (NIH), and the Office of Research in Women's Health at the NIH convened a workshop to explore how male and female biologic and physiologic characteristics affect musculoskeletal health. This issue of the *Orthopedic Clinics of North America* picks up where the workshop left off, extending the discussion of clinical topics across the broad spectrum of musculoskeletal health. This article serves as a prelude and introduction to the issue and provides a synopsis of the workshop findings.

Upper Extremity: Emphasis on Frozen Shoulder 531
Monique A. Sheridan and Jo A. Hannafin

> Adhesive capsulitis, or frozen shoulder syndrome, is a condition characterized by gradual loss of active and passive glenohumeral motion. The etiology of adhesive capsulitis is unknown. Treatment methods include supervised benign neglect, physical therapy, intra-articular corticosteroid injections, closed manipulation under anesthesia, arthroscopic capsular release, and open surgical release. Approximately 70% of patients presenting with adhesive capsulitis are women; however, the role of sex in the etiology, development, and outcome of treatment for adhesive capsulitis remains unclear. Individualized treatment is necessary following thorough evaluation of patient symptoms and stage of the disease.

How Men and Women Are Affected by Osteoarthritis of the Hand 541
E. Anne Ouellette and Anna-Lena Makowski

> Factors other than age and genetics may play a role in explaining the onset of osteoarthritis of the hand. Genetic, physiologic, and anatomic differences in men and women cause the variable expressions of osteoarthritis. These different factors affect women's ability to modify osteoarthritis of the hand before and after its onset, although it is genetic. By maintaining normal weight, good health, and nutrition, one can diminish the genetic and multifactorial effects of osteoarthritis of the hand. Future research in genetics, polymorphism, anatomy, hormonal influences, association with other disease processes, and multifactorial issues will clarify these relationships. Additional studies are needed to investigate the outcomes of gender-specific treatments, joint replacement surgery, and other interventions for osteoarthritis of the hand.

Sexual Dimorphism in Degenerative Disorders of the Spine 549
Neil A. Manson, Edward J. Goldberg, and Gunnar B. J. Andersson

> Sexual dimorphism is evident during formation, growth, and development of the spine. Pregnancy alters spine physiology and is a risk factor for back pain. The processes of aging and spinal degeneration adversely affect men and women slightly differently. Although degenerative changes are observed at similar rates in both sexes, women seem to be more susceptible to degenerative changes leading to instability and malalignment, such as degenerative spondylolisthesis. Men, however, suffer to a greater extent from structural deterioration, such as stenosis or disc degeneration. Surgical satisfaction is greater in men, which has been attributed to poorer preoperative function secondary to more advanced disease at time of surgery and lower patient expectations for clinical improvement, both observed in women.

Sexual Dimorphism in Adolescent Idiopathic Scoliosis 555
Cathleen L. Raggio

> Adolescent idiopathic scoliosis (AIS) is one of the orthopedic disorders in which clinical evidence of sexual dimorphism is most marked. Sexual dimorphism in spine growth, morphology, stiffness, curve pattern, and hormones may be the environment in which genetic factors combine to produce the phenotype of a scoliosis patient. These factors also may play a role in curve progression despite treatment and may help explain why some patients' curves never change and others are recalcitrant to nonoperative treatments. There are important differences in male and female AIS that impact diagnosis, treatment, and outcomes.

Osteoarthritis of the Hip and Knee: Sex and Gender Differences 559
Mary I. O'Connor

> Osteoarthritis of the hip and knee is a leading cause of functional disability and compromised quality of life in older patients and a significant public health issue. Emerging research shows sex and gender differences in osteoarthritis which, to date, may not be appreciated by the orthopedic community. This article discusses sex and gender differences in osteoarthritis with a focus on disease involving the hip and knee. Understanding what we know (and do not know) about sex and gender differences in this disorder is critical to improving quality of care for our patients.

Sexual Dimorphism of the Foot and Ankle 569
Kathryn O'Connor, Gwynne Bragdon, and Judith F. Baumhauer

> Lower extremity musculoskeletal injuries are extremely common. Sports-related sex differences, in addition to osteoporosis issues, have raised the level of social awareness that women's health care issues may be different than those of their male counterparts.

Traditional research investigation for the foot and ankle is focused on shoe style differences and the effect that these shoes have had on foot pain and injury (eg, bunion, lesser toe malalignment). In addition to the extrinsic factor of footwear, intrinsic factors, such as foot structure, ligamentous laxity, muscle strength, and proprioception, predispose individuals to injury. This article reviews the literature to examine the intrinsic and extrinsic differences between men and women in relationship to the foot and ankle and explores, where available, the influence that these factors have on injury.

Female Athletic Triad and Stress Fractures 575
David Feingold and Sharon L. Hame

Stress fractures are a common occurrence in athletes, and the incidence of stress fractures in female Division 1 collegiate athletes is double that of men. Hormonal influences on bone and bone morphology may influence the risk for fracture. A high level of suspicion and special imaging procedures allow for accurate diagnosis of these fractures. In stress fractures that are associated with the female athlete triad, addressing the three aspects of the triad—eating disorders, amenorrhea, and osteoporosis—are critical for successful treatment. Preparticipation screening for the presence of the signs of the female athlete triad by monitoring weight, energy level, menstrual cycles, and bone mineral density may help to prevent the occurrence of stress fractures in this population.

Anterior Cruciate Ligament Biology and Its Relationship to Injury Forces 585
James R. Slauterbeck, John R. Hickox, Bruce Beynnon, and Daniel M. Hardy

Anterior cruciate ligament injury is determined by two variables: the ultimate failure load of the ligament and the mechanical load applied to the ligament. All factors that contribute to anterior cruciate ligament injury must do so by affecting one or both of these two basic variables. Some factors, such as sex hormones and tissue remodeling, have a multifaceted effect on the failure load of the anterior cruciate ligament and the magnitude of the load applied to it. The model also illustrates the potentially profound effects that sex hormones and tissue remodeling likely have on female susceptibility to anterior cruciate ligament injuries.

Dimorphism and Patellofemoral Disorders 593
Elizabeth A. Arendt

Sex is defined as the classification of living things according to their chromosomal compliment. Gender is defined as a person's self-representation as male or female or how social institutions respond to that person on the basis of his or her gender presentation. One frequently divides the topic of dimorphism into the biologic response inherent in their sex and the environmental response that might be better termed "gender differences." Clinicians have anecdotally agreed for years that patellofemoral disorders are more common in women. Given the difficulty in classifying patellofemoral disorders, literature support for this assumption is meager. For the purposes of this article we divide patellofemoral disorders into three categories: patellofemoral pain, patellofemoral instability, and patellofemoral arthritis. Possible sex differences in these disorders are reviewed.

Osteoporosis: Differences and Similarities in Male and Female Patients 601
Joseph Michael Lane, Alana Carey Serota, and Bradley Raphael

Osteoporosis is associated with decreased bone strength as a consequence of decreased bone density and altered quality. It is a result of a disruption of balance between bone breakdown and bone formation, caused by increased bone resorption by osteoclasts or deficient bone replacement by osteoblasts. The "silent thief" affects women and men; without appropriate screening, one's first awareness of the disease is a fracture. It results in increased mortality and significant morbidity. In the last decade, great strides have been made in defining the diagnosis and establishing effective modes of treatment for this disorder. Our current state of knowledge indicates that although this disease affects both sexes, there are clear differences that have clinical importance.

Hip Fracture and Its Consequences: Differences Between Men and Women 611
Denise L. Orwig, Julia Chan, and Jay Magaziner

This article describes the state of knowledge regarding gender differences with respect to hip fracture and its subsequent outcomes. Most of the work to date investigating hip fracture patients has been done with women, yet some evidence from a few studies with a significant number of male hip fracture patients and from nonfracture samples suggests that women and men may be different at the time of fracture and will have a different course of recovery.

Sexual Dimorphism in Stroke 623
Mary Ann Keenan

Stroke is a leading cause of death and serious, long-term disability. Studies evaluating differences between men and women are lacking. Significant differences exist between men and women in terms of risk factors and susceptibility to stroke. Women are less likely to have diagnostic studies performed to evaluate their risk for stroke, and they have a higher mortality following acute stroke. Women however, have a higher rate of arterial recanalization after intravenous tissue plasminogen activator used for the treatment of acute stroke. The data comparing the effectiveness of treatments for prevention of recurrent stroke between men and women is sparse. There have not been any studies comparing results of treatment of musculoskeletal impairments in men and women after stroke.

Sex Differences in Autoimmune Disease 629
Michael D. Lockshin

Many, but not all, autoimmune diseases primarily affect women. In humans, severity of illness does not differ between men and women. Men and women respond similarly to infection and vaccination, which suggests that the intrinsic differences in immune response between the sexes do not account for differences in disease frequency. In autoimmune-like illnesses caused by recognized environmental agents, sex discrepancy is usually explained by differences in exposure. Endogenous hormones are not a likely explanation for sex discrepancy; hormones could have an effect if the effect is a threshold rather than quantitative. X and Y chromosomal differences have not been studied in depth. Other possibilities to explain sex discrepancy include chronobiologic differences and various other biologies, such as pregnancy and menstruation, in which men differ from women.

Sex-Based Centers of Care: A Look to the Future 635
Kimberly J. Templeton

Although sex-based centers of care have played a critical historic role in improving the health of their constituents and drawing attention to the different health care needs of men and women, it is time to challenge them to do an even better job. There are now overwhelming data showing that men and women are profoundly different at the molecular and cellular level in virtually all aspects of musculoskeletal health and disease, but the clinical implications of these differences have generally been unexplored. Sex-based health centers of care can play a critical role in exploring these differences and, in doing so, reduce disability and enhance quality of life in our growing population of senior men and women.

Index 639

FORTHCOMING ISSUES

January 2007
Vascularized Bone Grafting in Orthopedic Surgery
Alexander Y. Shin, MD
Steven L. Moran, MD, *Guest Editors*

April 2007
Wrist Trauma
Steven Papp, MD
Allan Giachino, MD, *Guest Editors*

July 2007
Minimally Invasive Spine Surgery
Dino Samartzis, MD
Francis H. Shen, MD
D. Greg Anderson, MD, *Guest Editors*

RECENT ISSUES

July 2006
Advances in Musculoskeletal Imaging
Peter L. Munk, MD, CM, FRCPC
Bassam Masri, MD, FRCSC, *Guest Editors*

April 2006
The Pediatric Hip
James T. Guille, MD, *Guest Editor*

January 2006
Oncology
Rakesh Donthineni, MD, *Guest Editor*

The Clinics are now available online!

Access your subscription at:
http://www.theclinics.com

Preface

Laura L. Tosi, MD   Letha Y. Griffin, MD, PhD   Mary I. O'Connor, MD
Guest Editors

There's no denying it: Males and females differ in every body system. Yet discussions of these basic differences are often impeded by the absence of a single, three-letter word—"sex." Using the "S" word seems to cause discomfort, even though "sex," not "gender," is the more scientifically correct term. According to the Institute of Medicine, *sex* is "the classification of living things, generally male or female, according to their reproductive organs and functions assigned by the chromosomal complement," whereas *gender* is " a person's self-presentation as male or female, or how that person is responded to by social institutions on the basis of the individual's gender presentation" [1]. In some cultures, for example, gender (ie, being perceived as female) has a critical impact on matters such as access to nutrition and exercise. Thus "sex" determines the inner workings of an individual's growth and development, whereas "gender" encompasses the environment in which an individual grows and lives.

The science supporting the need to recognize sexual dimorphism in health and disease is now compelling. Recent studies in mice demonstrate that thousands of genes, affecting a broad array of tissues, including liver, fat, muscle, and brain, evince sexual dimorphism. These genes exhibit highly tissue-specific patterns of expression. At the same time, the medical community, as well as the popular press, has slowly started to recognize the importance of sexual dimorphism in clinical care, particularly in heart disease. However, little attention has been paid to sexual dimorphism in musculoskeletal disorders, except in the areas of osteoporosis and anterior cruciate ligament injuries of the knee.

This issue of the *Orthopedic Clinics of North America* documents much of what we currently know about sexual dimorphism in musculoskeletal health, and draws attention to the critical need for more sex-specific research to ensure that the prevention, diagnosis, and treatment of musculoskeletal disorders are appropriate to the sometimes different needs of men and women. Currently, there are efforts to provide more appropriately designed implants for total knee replacement in women, for example, now that it is clear that women's knees are different, not just smaller than those of men. Other researchers are exploring osteoporosis care from the standpoint of reducing the higher mortality rates in men with hip fractures, as well as determining whether osteoporosis drugs that are effective in women are also effective in men. Clearly, the important clinical question must be: Under what clinical circumstances does sex influence how musculoskeletal care should be delivered?

This is a quality-of-care issue. The articles in this issue call us to action: to improve quality of care for all of our patients, both men and women. Clinicians have always organized their care by patient age and other salient factors; now the data suggest that discrimination by sex is just as important. Sex—so often the focus of excessive attention in other sectors of our society—turns out to matter in an arena in which we heretofore gave it too little heed: musculoskeletal health. It's time for a change.

Laura L. Tosi, MD
Division of Orthopaedic Surgery
Children's National Medical Center
111 Michigan Avenue, NW
Washington, DC 20010, USA

E-mail address: LTOSI@cnmc.org

Letha Y. Griffin, MD, PhD
Peachtree Orthopedic Clinic
2045 Peachtree Road NE, Suite 700
Atlanta, GA 30309, USA

E-mail address: lethagriff@aol.com

Mary I. O'Connor, MD
Mayo Clinic Jacksonville
4500 San Pablo Road
Jacksonville, FL 32224, USA

E-mail address: oconnor.mary@mayo.edu

Reference

[1] Wizemann TM, Pardue ML, editors. Institute of Medicine (US) Committee on Understanding the Biology of Sex and Gender Differences. Exploring the biological contributions to human health: does sex matter? Washington, DC: National Academy Press; 2001.

Past and Future: Sex and Gender in Health Research, the Aging Experience, and Implications for Musculoskeletal Health

Vivian W. Pinn, MD

Office of Research on Women's Health, National Institutes of Health, Department of Health and Human Services, 6707 Democracy Boulevard, Suite 400, MSC 5484, Bethesda, MD 20892-5484, USA

The past 2 decades have witnessed widespread efforts among scientists, advocates, and health care providers of diverse disciplines and specialties to redefine what is meant by "women's health." These efforts have included clarifying the distinction between "sex" and "gender," and developing an interdisciplinary approach to women's health research and its implications for clinical practice. As a result, we now understand more about the interrelationships among genetic makeup, biologic aspects, socioeconomic factors, health care delivery, and personal lifestyle behaviors of individuals, and how these independently or together contribute to the health status of women and men across their lifespan.

Through research, we have made progress in preventing premature death from many major life-threatening diseases. As the average life span increases, we have the further challenge of preserving physical and mental abilities into advanced age and sustaining a good quality of life. This is particularly an issue for women, given their greater life expectancy and increasing numbers in advanced age groups. Musculoskeletal health has a key role in quality of life at all ages. Understanding how musculoskeletal health contributes to wellness or disability in later life is essential to addressing the needs of our increasingly aging population. Understanding the differences and similarities between and among women and men in the causes and expression of musculoskeletal development and health is critical to tailoring health interventions appropriately to provide sex- and gender-appropriate prevention strategies and the best quality of health care.

Although recent efforts of advocates began with a focus on issues of women's health, the need emerged to ensure that research addresses gaps in knowledge related to differences, or similarities, between health and disease processes in women and men and their implications for clinical practice, including diagnostics, therapeutic regimens, and preventive approaches.

Evolution and current status of women's health research

As recently as 20 years ago, women's health research focused primarily on reproductive health. Traditionally, clinical research on conditions that affect women and men routinely did not include women; the medical literature provided few analyses of differences between women and men, even if women were included in the study populations. That approach is undergoing a dramatic change. Today, there is greater awareness of the importance of exploring sex and gender[1] differences and

E-mail address: pinn@od1th1.od.nih.gov

---

[1] The Institute of Medicine 2001 report *Exploring the Biological Contributions to Human Health: Does Sex Matter?* recognizes the need to establish a clear definition of the terms "sex" and "gender" and a consistent use of these terms in the medical literature. The report recommended that the term "sex" be used when differences are primarily biologic in origin and may be genetic or phenotypic (genetic or physiologic characteristics of being male or female), and that "gender" be used when referring to responses to social and cultural influences based on sex.

similarities between women and men in outcomes of research studies, especially because the appreciation of women's health now includes the expanded concept of body health beyond the reproductive system and the reproductive years of life. Of major interest are genetic, biologic, behavioral, and socioeconomic factors that result in health disparities between men and women as well as between different population groups.

This change in research focus has not happened by accident. It arose, in large part, because of leadership in the 1980s in the women's health movement that questioned the lack of participation of women in research studies of conditions that affect men and women, so that medical care for women would be based on norms defined by women, rather than on the assumption that the results of studies in men are applicable to women. That movement essentially led to the current attention on defining sex and gender factors in health and disease and to the establishment in 1990 of the Office of Research on Women's Health (ORWH) at the National Institutes of Health (NIH). ORWH serves as a focal point for policy and research leadership and collaboration across the NIH for women's health research. From its position within the Office of the Director of the NIH, ORWH works in concert with the NIH's institutes and centers and the biomedical, advocacy, and health care communities to formulate and implement an agenda for research on women's health issues, including a focus on sex and gender contributions to the health of women and men.

NIH published its first agenda for research on women's health issues in 1992. Since then, ORWH has updated that agenda annually and revised its priorities accordingly. In a major outreach effort in 1996 and 1997, ORWH convened a series of national scientific workshops and public hearings to identify research priorities for a revised agenda. More than 1500 basic, clinical, and social scientists, health care provides, educators, and individuals from community organizations and advocacy groups formulated recommendations about areas of continuing or emerging scientific interest in need of further research and strategies, including research on bone and musculoskeletal diseases. The result of this extended effort, *Agenda for Research on Women's Health for the 21st Century: A Report of the Task Force on the NIH Women's Health Research Agenda for the 21st Century* [1] included not only sex- and gender-specific priorities for women, but also priorities for investigations of those health conditions that affect men and women for which more data are needed to determine how these disorders or diseases may differ between the sexes. A major focus of current research is to design and implement basic and clinical studies on conditions that affect women and men in such a way that analyses by sex/gender can be conducted to determine if differences do exist for women, thereby providing information that can be used in gender-specific health care [2].

In 1994, NIH expanded its policy on inclusion of women in clinical research studies to meet the mandate of the NIH Revitalization Act of 1993 that women and minorities be included in clinical research studies that are funded by the NIH [3]. This revised policy, based on federal law, has been instrumental in assisting NIH to implement strengthened requirements for including women in clinical research of conditions that affect women and men, and to call for an analysis of results for differences based on sex and gender. The intent is to ensure that scientific norms for health, disease, treatments, and other medical interventions are applicable to all populations (men and women, and diverse racial/ethnic groups), based on scientific evidence established through research. Although NIH requires that final progress reports of clinical research must include gender analyses, NIH does not have the authority to require that medical journals include such analyses in their published manuscripts. That remains an issue to be addressed by the editorial staff of individual journals; however, efforts to emphasize the importance of clinical differences in responses to interventions or health strategies in the improved care of women and men should help emphasize to authors and editors the importance of including such data. The US Food and Drug Administration (FDA) has put in place guidelines for including women and minorities in drug trials, and it requires reporting the analyses of the effects of the drugs for both sexes and for major racial groups as a condition for approval and marketing of new medications [4]. With the NIH and FDA emphasizing the importance of analyses of research data by sex and gender, a joint effort was undertaken to provide an online course, *The Science of Sex and Gender in Human Health* [5]. This course is designed to give a basic scientific understanding of the major physiologic differences between the sexes, the influence that those differences have on illness and health outcomes, and the implications for policy, medical research, and health care.

The NIH policy for inclusion is applicable only to clinical research. Therefore, the online course also is intended to influence investigators about the need to consider potential differences based on sex in the design and interpretation of basic science studies. The Institute of Medicine publication, *Exploring the Biological Contributions to Human Health: Does Sex Matter?* [6], has become instrumental in providing the scientific reasoning behind encouraging researchers to determine sex/gender factors related to basic biologic studies: "An additional and more general reason for studying differences between the sexes is that these differences, like other forms of biological variation, can offer important insights into underlying biological mechanisms." As expressed by Federman [7], "Differences between the sexes pervade all clinical experience in medicine."

Studies to address the causes, treatment, and prevention of disparities between populations of women and men, including their subpopulations, may require single-sex composition and may be designed to address the interaction between an individual's genetic and biologic dispositions, environment, personal health behaviors, racial/ethnic/cultural attributes, access to health care, and many other aspects that may contribute to differences in health status or outcomes between different populations [8].

The current priorities for research on women's health reflect the evolution of women's health research from a focus on the expanded life span concepts of women, to a focus on sex and gender determinants as the basis for sex- and gender-appropriate medical care in the twenty-first century [9]. Research priorities for stimulating new studies now emphasize chronic and preventable illnesses and complex multisystemic conditions, in addition to focusing on specific diseases and conditions across the life span. Recommended ORWH priorities also indicate that gender and sex are important considerations in most areas of research, including psychologic, social, and behavioral studies; consideration of these variables is critical to the accurate interpretation and validation of research affecting women's health, because these variables determine how similar or different health or disease processes may be between women or between men and women.

The expanded concept of what constitutes women's health also has led to the recognition that research priorities and health care must be less fragmented and more comprehensive and interdisciplinary. The concept of interdisciplinary research, as a foundation for comprehensive medical care, has been reinforced and enhanced through ORWH programs [10] and through NIH Roadmap initiatives [11,12] for novel interdisciplinary training and clinical approaches. There is also a clear need for research that will—without slighting the unique needs of women—pay equal attention to age-related issues in men and women.

Early research progress on women's health has deepened our understanding of sex and gender differences and similarities in the etiology, diagnosis, progression, treatment, health outcomes, and prevention of many conditions that affect women and men. For example, there has been progress in identifying and understanding the role of various metabolic enzymes in causing sex differences in pharmacodynamics. Studies using combinations of neuroanatomic and behavioral techniques have identified sex differences in the manifestations of brain disorders and the influence of reproductive endocrine disorders in women and men. Many researchers now examine sex and gender, socioeconomic, and racial and ethnic differences when studying the incidence, mortality, and prevention of diseases. Research increasingly is defining the effects of hormones on the biologic, genetic, environmental, and behavioral aspects of health. Advances in musculoskeletal health also must include a better understanding of the effects of genetics and genetic mutations relevant to conditions of concern, such as arthritis and of cell physiology, matrix factors, and coenzyme factors. Improvements have been made in recognizing sex and gender factors in joint replacement surgery and physical therapy, but there are many other areas for which medical questions remain about potential differences in clinical approaches for musculoskeletal conditions in women and men.

Future directions for musculoskeletal research

The musculoskeletal system provides a striking means of illustrating the value of a collaborative, comprehensive approach that integrates basic and clinical research with preventive and behavioral research. The benefits of research that addresses the continuity of musculoskeletal issues throughout the life span are incalculable. Musculoskeletal health is built, maintained, and protected in the early decades of life, and is a major influence on quality of life in later years. Research is needed to elucidate genetic, molecular, cellular mechanisms, and pathways in the healthy development of the system and in pathogenesis. Are these factors and processes different or similar in women and men? For example, we know that arthritis is more common in women than in men. Is this difference due to a woman's genetic makeup and hormones? Do structural and biochemical factors predispose women to develop arthritis? Among women and men, are certain subpopulations more susceptible—whether because of genetics, lifestyle, or socioeconomic factors—to musculoskeletal diseases and disabilities? How does lifestyle strengthen or weaken the musculoskeletal system? As females are increasingly engaging in strenuous physical activity, military training, and professional sports, is the medical community fully informed about how sex-based factors may influence injuries or treatment strategies?

Longevity and the challenges of quality of life and quality of health care

Now that the average life span in the United States is growing longer, addressing the challenges of providing quality health care to the aging population and of motivating this population to practice self-care behaviors that will prevent disability in later life has become essential. Four key factors characterize our aging population [13]:

Men and women are living longer; however, on average, women continue to live longer than do men.

A large number of our nation's citizens are disabled during the aging years; the percentage of women that has disabling conditions is greater than that of men.

Much of age-related disability is associated with conditions of the musculoskeletal system.

Multiple conditions may make individuals susceptible throughout life to specific diseases or disabilities may begin early in life and become manifest later. Of these conditions, lifestyle is among the most important.

Increased life expectancy

In 2004, life expectancy at birth for the United States population reached a record high of 77.9 years—an increase of 0.4 years relative to 2003. The age-adjusted death rate declined significantly for 10 of the 15 leading causes of death. Long-term trends continued in decreased deaths from heart disease, cancer, and stroke, which remain the three leading causes of mortality [13]. As a result, the older population in 2030 is projected to be twice as large as in 2000, increasing from 35 million to 72 million, and represent nearly 20% of the total United States population at the latter date [14].

As both men and women live longer lives on average, women still live longer, on average, than do men. The difference between life expectancy in 2004 for men and women was 5.2 years (75.2 and 80.4 years, respectively), the smallest such difference since 1946 (Fig. 1) [13]. The number of

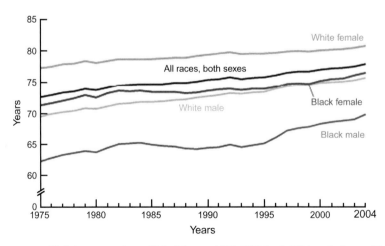

Fig. 1. Life expectancy at birth by race and sex: United States, 1975–2003 final, 2004 preliminary. (*From* Miniño AM, Heron M, Smith BL. Deaths: preliminary data for 2004. Health E-Stats. Released April 19, 2006. Available at: http://www.cdc.gov/nchs/products/pubs/pubd/hestats/prelimdeaths04/preliminarydeaths04.htm.)

centenarians increased from about 37,000 in 1990 to more than 50,000 in 2000. About 80% of centenarians are women (Fig. 2) [14].

Disability during the older years

Census 2000 counted about 14 million civilian noninstitutionalized older people with some type of disability (Fig. 3). Older women were more likely than were older men to experience disability (43% and 40% respectively) [14].

> In 2004, among adults aged 55 years and older, men and women were equally likely to be in fair or poor health: 23.2% were in fair or poor health—ranging from 19.6% of adults aged 55 to 64 years to 33.6% of adults aged 85 years and older.
>
> Non-Hispanic black adults and Hispanic adults were more likely than were non-Hispanic white adults and non-Hispanic Asian adults to be in fair or poor health, in all but the oldest age group.
>
> Poor adults were more likely than were adults who were not poor to be in fair or poor health; the greatest differences in health status by poverty status were for adults aged 55 to 64 years.

Disabilities can manifest themselves in the simplest daily activities that robust people take for granted, as illustrated by the following 2004 data for older age groups:

- About 1 in 4 adults aged 55 years and older had difficulty walking a quarter mile, ranging from 16.5% of adults aged 55 to 64 years to more than 50% of adults aged 85 and over.
- About 1 in 5 adults aged 55 years and older had difficulty walking up 10 steps. Adults aged 85 years and older (46.2%) were nearly four times as likely as adults aged 55 to 64 years (12.9%) to have difficulty with this activity.
- About 1 in 4 adults aged 55 years and older had difficulty standing for 2 hours, and prevalence of this difficulty increased with age.
- About 1 in 5 adults aged 55 years and older had difficulty pushing or pulling large objects, with rates for those aged 85 years and older (46.6%) triple those of adults aged 55 to 64 years (15.5%).
- About 1 in 10 adults aged 55 years and older had difficulty shopping (12.4%) or socializing. Rates increased gradually between ages 55 to 64 years and ages 75 to 84 years and then doubled for the group aged 85 years and older. About one third of adults in the oldest age group had difficulty shopping or socializing (29.2%).
- Across the activities studied, women were more likely than were men to have difficulty with physical and social activities, with the largest differences noted in the age groups aged 65 years and older.

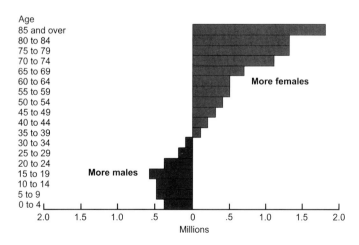

Fig. 2. Difference between male and female populations by age: 2000. The reference population for these data is the resident population. (*From* Wan H, Sengupta M, Velkoff VA, et al. US Census Bureau, Current Population reports, P23–209, 65+ in the United States: 2005. Washington DC: US Government Printing Office; 2005. p. 23.)

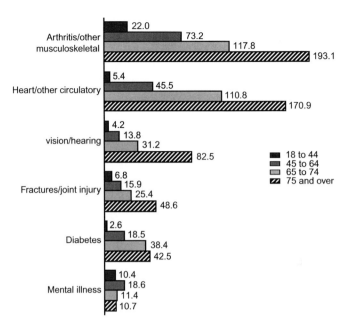

Fig. 3. Selected chronic health conditions causing limitation of activity among adults by age: 1998 to 2000. Number of people with limitation of activity caused by selected chronic health condition per 1000 population. The reference population for these data is the civilian noninstitutionalized population. (*From* Wan H, Sengupta M, Velkoff VA, et al. US Census Bureau, Current Population reports, P23–209, 65+ in the United States: 2005. Washington DC: US Government Printing Office; 2005. p. 55.)

Age-related disabilities of the musculoskeletal system

Chronic diseases and impairments, which are among the leading causes of disability in older people, can affect quality of life negatively, lead to a decline in independent living, and impose an economic burden [15,16]. Multiple conditions may make individuals susceptible throughout life to specific diseases or disabilities, although the onset of a condition may become manifest only later in life. About 80% of seniors have at least one chronic health condition and 50% have at least two [17]. Arthritis, hypertension, heart disease, diabetes, and respiratory disorders are among the leading causes of activity limitations in older people [14].

Arthritis

Arthritis encompasses more than 100 diseases and conditions that affect joints, surrounding tissues, and other connective tissues. It is the leading cause of disability in the United States.

- 42.7 million adults, or just over one in five adults in the United States have doctor-diagnosed arthritis [18].
- Arthritis is more common among older people than younger people and is more common in women of all ages than among men [14,19,20].
- Arthritis affects persons of all races and ethnic groups [14].
- In 1998 to 2000, 19.3% of people aged 75 years and older and 11.8% of people aged 65 to 74 years had activity limitations that were caused by arthritis and other musculoskeletal conditions, compared with 2.2% of people aged 18 to 44 years [14].

Osteoporosis

Osteoporosis, or loss of mass and quality of bones, is a second common chronic ailment among older people. Osteoporosis reduces bone density and raises the risk for potentially disabling fractures [16,19,22]. Hip fractures are particularly disabling and also may increase the subsequent risk for mortality [14,21,23].

As reported in the Surgeon General's report, *Bone Health and Osteoporosis* [22]:

> An estimated 10 million Americans older than 50 years have osteoporosis, which makes it

the most common bone disease. Another 34 million are at risk.

Each year an estimated 1.5 million people suffer an osteoporotic-related fracture, an event that often leads to a downward spiral in physical and mental health.

Twenty percent of senior citizens who suffer a hip fracture die within 1 year.

Fifty percent of women older than 50 years will have an osteoporosis-related fracture in their lifetime, with the risk for fracture increasing with age.

Due primarily to the aging of the population and the previous lack of focus on bone health, the number of hip fractures in the United States could double, or even triple, by 2020.

Although most health research has focused on men, the reverse is true with respect to osteoporosis. Nine out of every 10 people who have osteoporosis in the United States are women. Not surprisingly, until recently, most studies about osteoporosis have been done with white women, so there is less knowledge about this condition in men than in women. Meanwhile, the incidence of osteoporosis is increasing in men and women (Fig. 4), especially because of increased longevity.

This fact has led to such clinical questions, as "What are effects of differential bone size between sexes on risk for fractures?" and "Should the same bone mineral density criteria for diagnosis and for fracture risk determined in studies on women be used for men?" [24].

Impact of physical activity on musculoskeletal disease

Evidence increasingly supports the positive link between physical activity and the onset and progression of musculoskeletal disease [25]. In adults, physical activity decreases the risk for cardiovascular diseases, diabetes, musculoskeletal problems, and cancer, and increases strength, physical functioning, and longevity [19,25–27]. Older women (26.1%) are more likely to be inactive than are older men (17.7%) [25,28]. Among older men and women who are active, older women are less likely to have high overall activity levels (18.2% of older men and 13.1% of older women).

Leisure-time physical activity

In 2004, about one half of adults aged 55 years and older (51.7%) engaged—regularly or irregularly—in at least some light, moderate, or

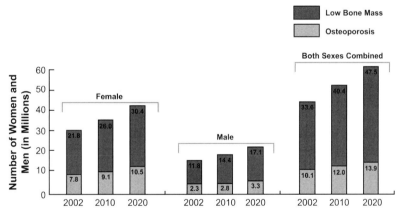

Fig. 4. Projected prevalence of osteoporosis and/or low bone mass of the hip in women, men, and both sexes, 50 years of age or older. Note: The National Health and Nutrition Examination Survey (NHANES) is conducted by the National Center for Health Statistics, a part of the Centers for Disease Control and Prevention. This survey is conducted on a nationally representative sample of Americans. As a part of NHANES, bone mineral density of the hip was measured in 14,646 men and women over 20 years of age throughout the United States from 1988 until 1994. These values were compared with the WHO definitions to derive the percentage of individuals over the age of 50 who have osteoporosis and low bone mass. These percentages were then applied to the total population of men and women over the age of 50 to estimate the absolute number of males and females in the United States with osteoporosis and low bone mass. Projections for 2010 and 2020 are based on population forecasts for these years; they are significantly higher than current figures due to both the expected growth in the overall population and the expected aging of the population. (*From* US Department of Health and Human Services. Bone health and osteoporosis: a report of the Surgeon General. Rockville, MD: US Department of Health and Human Services. Office of the Surgeon General; 2004. p. 76.)

vigorous leisure-time physical activity. This ranged from more than one half of adults aged 55 to 64 years to about one fourth of adults aged 85 years and older. Differences between men and women in rates of participation in any leisure-time physical activity varied by age. Among adults aged 55 to 64 years, rates of participation were about the same for men (58.1%) and women (57.1%). Among adults aged 65 years and older, men (52.6%) were more likely than were women (43.3%) to participate in leisure-time physical activities [28].

Regular leisure-time physical activity

Overall, about one in four adults aged 55 years and older (24.7%) engaged in regular leisure-time physical activity. The prevalence was about the same for adults aged 55 to 64 years (28.1%) and adults aged 65 to 74 years (26.9%), but it was markedly lower among adults aged 75 to 84 years (18.7%) and those aged 85 years and older (8.2%) [28].

Men were more likely than were women to engage in regular leisure-time physical activity, with the greatest differences found in the oldest age groups. Among adults aged 55 to 64 years, men (29.9%) were more likely than were women (26.5%) to engage in regular leisure-time physical activity. Among adults aged 75 to 84 years, men (24.2%) were about 1.5 times as likely as were women (15.0%) to engage in regular leisure-time physical activity. Among adults aged 85 years and older, men (11.6%) were almost twice as likely as were women (6.5%) to engage in regular leisure-time physical activity.

Strengthening activities

Rates of participation in activities that are designed to strengthen muscles were low among adults in all age groups aged 55 years and older, ranging from 17.2% of adults aged 55 to 64 years to 7.2% of adults aged 85 years and older. Adults who were not poor were two to three times as likely as were poor adults to engage in strengthening activities. Among adults aged 55 to 64 years, about 1 in 5 adults who were not poor (21.4%) engaged in strengthening activities compared with fewer than 1 in 10 poor adults (7.9%). Among adults aged 65 years and older, 16.5% of those who were not poor did strengthening exercises compared with 6.8% of poor adults. Among adults aged 85 years and older, 12.1% of adults who were not poor did strengthening activities compared with 4.0% of poor adults [28].

Summary

The statistics about common musculoskeletal disorders describe a few of the many health conditions that affect men and women. For such disorders and conditions, there are differences in incidence, predisposition, and therapeutic and preventive strategies for managing them. Although we have made progress in women's health research, many challenges remain, including those related to conditions and diseases of the musculoskeletal system that may affect women and men differentially. Research is needed to identify genetic, hormonal, environmental, and societal factors that contribute to these sex and gender differences, and to understand when appropriate clinical applications should differ or be the same. For example, many of these conditions are associated with pain; therefore, sex differences in response to the analgesic effects of different pharmacologic agents may be clinically relevant.

The health of the musculoskeletal system is critical in the ability of men and women of all ages to prevent life-threatening and life-debilitating diseases and conditions and to maintain an active, healthy lifestyle into the elderly years. As collaborative, interdisciplinary research yields treatment options for musculoskeletal diseases and conditions, health care practitioners and public health professionals must be informed about the role of sex and gender factors in health and disease so that they can provide the most appropriate medical care and communicate to their patients and the public what they can do to promote their musculoskeletal wellness throughout life.

References

[1] Office of Research on Women's Health. Agenda for Research on Women's Health: A Report of the Task Force on the NIH Women's Health Research Agenda for the 21st Century. Available at: http://orwh.od.nih.gov. Accessed April 20, 2006.
[2] Pinn VW. Sex and gender factors in medical studies: implications for health and clinical practice. JAMA 2003;289:397–400.
[3] US Department of Health and Human Services, National Institutes of Health. "NIH Guidelines on the Inclusion of women and Minorities as Subjects in Clinical Research." Federal Register 59, 14508–13 (28 March 1994).
[4] Food and Drug Administration. "Guidelines for the study and evaluation of gender differences in the clinical evaluation of drugs." Federal Register 58, 39406 (23 July 1993).

[5] The science of sex and gender in human health: online course site. Available at: http://sexandgendercourse.od.nih.gov. Accessed April 20, 2006.
[6] Wizemann TM, Pardue M-L, editors. Exploring the biological contributions to human health: does sex matter? Washington, DC: National Academy Press; 2001.
[7] Federman DD. The biology of human sex differences. N Engl J Med 2006;354:1507–14.
[8] Pinn VW. Challenges in advancing women's health in the 21st century: perspectives from the Office of Research on Women's Health. Ethn Dis 2001;11:177–80.
[9] FY2006 NIH research priorities for women's health. http://orwh.od.nih.gov/research/FY06research_priorities.pdf. Accessed April 20, 2006.
[10] Building interdisciplinary research careers in Women's Health program description. Available at: http://www4.od.nih.gov/orwh/bircwhmenu.html.
[11] NIH Roadmap. Available at: http://nihroadmap.nih.gov/index.asp; http://nihroadmap.nih.gov/interdisciplinary/.
[12] Elias A. Zerhouni. US biomedical research: basic, translational, and clinical sciences. JAMA 2005;294:1352–8.
[13] Miniño AM, Heron M, Smith BL. Deaths: preliminary data for 2004. Health E-Stats. Released April 19, 2006. Available at: http://www.cdc.gov/nchs/products/pubs/pubd/hestats/prelimdeaths04/. Accessed April 19, 2006.
[14] Wan H, Sengupta M, Velkoff VA, et al. US Census Bureau, Current Population reports, P23–209, 65+ in the United States: 2005. Washington DC: US Government Printing Office; 2005.
[15] Centers for Disease Control and Prevention. Unrealized prevention opportunities: reducing the health and economic burden of chronic disease. Department of Health and Human Services, 1997.
[16] National Center for Health Statistics. National health interview survey, selected years. Health, United States, with health and aging chartbook, 1999. Centers for Disease Control and Prevention, National Center for Health Statistics, Department of Health and Human Services; 1999. Publication No. 99-1232. Available at: http://www.cdc.gov/nchs/nhis.htm.
[17] Centers for Disease Control and Prevention. Healthy aging: preventing disease and improving quality of life among older Americans. In: At a glance. Department of Health and Human Services; 2003.
[18] Quick stats. Available at: http://cdc.gov/arthritis/pressroom/idex.htm.
[19] Blackman, Donald K, Kamimoto LA, et al. Overview: surveillance for selected public health indicators affecting older adults-United States. Morb Mortal Wkly Rep Surveill Summ 1999;48(SS08):1–6.
[20] Centers for Disease Control and Prevention. Targeting arthritis: the nation's leading cause of disability. In: At a glance. Department of Health and Human Services; 2003.
[21] Magaziner J, Lydick E, Hawkes W, et al. Excess mortality attributable to hip fracture in white women aged 70 years and older. Am J Public Health 1997;87(10):1630–6.
[22] US Department of Health and Human Services. Bone health and osteoporosis: a report of the Surgeon General. Rockville, MD: US Department of Health and Human Services. Office of the Surgeon General; 2004.
[23] Wolinsky, Fredric D, John Fitzgerald J, et al. The effect of hip fracture on mortality, hospitalization, and functional status: a prospective study. Am J Public Health 1997;87(3)398–403.
[24] Miller PD. The clinical application of gender-based databases for the diagnosis of osteoporosis and fracture risk prediction. In: Legato MJ, editor. Principles of gender specific medicine, vol. 2. Amsterdam: Elsevier Academic Press; 2004. p. 1033–42.
[25] Schoenborn CA, Adams PF, Barnes PM. Body weight status of adults: United States, 1997–98. Adv Data 2002 Sep 6;(330):1–15.
[26] Powell, Kenneth E, Thompson PD, et al. Physical activity and the incidence of coronary heart disease. Annu Rev Public Health 1987;8:253–87.
[27] Keysor JJ, Jette AM. Have we oversold the benefit of late-life exercise? J Gerontol A Biol Sci Med Sci 2001;56(7):M412–23.
[28] Schoenborn CA, Vickerie JL, Powerll-Griner E. Health characteristics of adults 55 years of age and over: United Sates, 2000–2003. DHHS Pub No. 2006-1250. Centers for Disease Control and Prevention. National Center for Health Statistics.

Does Sex Matter in Musculoskeletal Health? A Workshop Report

Laura L. Tosi, MD[a,*], Barbara D. Boyan, PhD[b], Adele L. Boskey, PhD[c]

[a]*Division of Orthopaedics, Children's National Medical Center, 111 Michigan Avenue NW, Washington, DC 20010, USA*
[b]*Department of Biomedical Engineering, Georgia Institute of Technology, 315 Ferst Road NW, Atlanta, GA 30332-0363, USA*
[c]*Hospital for Special Surgery, 535 East 70th Street, New York, NY 10021, USA*

We do not know why disease prevalence rates differ for men and women, or why the average life expectancies of women are substantially longer than those of men [1]. The past decade, however, has seen enormous progress toward understanding what the bases of these differences might be.

Only in the past generation has it been possible to explore at the genetic and cellular levels the differences in disease prevalence and the differences in morbidity and mortality between men and women. The completion of the Human Genome Project in 2003 now enables the elucidation of the genetic basis of disease. We can now make a persuasive case for the existence of innate and heretofore unexamined differences between men and women. Researchers now realize that every organ in the body, not just the reproductive system, responds differently on the basis of sex. These different responses result from chromosomes and from hormones. In other words, a male cell is not the same as a female cell, and sex chromosome–linked genes can be expressed in germ line and somatic cells.

Musculoskeletal medicine is one area in which the effects of sex have the greatest influence on treatment and outcome. Osteoarthritis, osteoporosis, spinal disorders, and fractures all occur more often in women than in men. Differences in injury mechanism, pain sensation, drug handling, and healing response have a biologic basis. Responses to surgery, anesthesia, medications, and rehabilitation also differ with sex. Yet, in most cases, we have based therapeutic modalities on men. Most clinicians are unaware of the major effect of inherent differences in biology at the cellular and molecular levels.

In April 2004, the American Academy of Orthopaedic Surgeons (AAOS), the National Institute of Arthritis and Musculoskeletal and Skin Diseases of the National Institutes of Health (NIH), and the Office of Research in Women's Health at the NIH convened a workshop to explore how male and female biologic and physiologic characteristics affect musculoskeletal health. Initiated by the AAOS Women's Health Issues Committee and cosponsored by the AAOS Committee on Research, the workshop took its lead from a 2001 Institute of Medicine (IOM) report, "Exploring the Biological Contributions to Human Health: Does Sex Matter?" [1,2]. That report emphasized four themes: every cell has a sex, sex begins in the womb, sex affects

---

Major portions of this material were previously published in Tosi LL, Boyan BD, Boskey AL. Does sex matter in musculoskeletal health? The influence of sex and gender on musculoskeletal health. J Bone Joint Surg Am 2005;87:163–47 and are republished here by permission of The Journal of Bone and Joint Surgery, Inc.

* Corresponding author.
E-mail address: ltosi@cnmc.org (L.L. Tosi).

behaviors and perception, and sex affects health.[1] Workshop participants sought to answer the following questions: Are there sex-based differences that are important to orthopedics? If so, how can they be identified, studied, and overcome?

A detailed description of that meeting and its recommendations is provided in the authors' earlier work, "Does Sex Matter in Musculoskeletal Health? The Influence of Sex and Gender on Musculoskeletal Health," published in the *Journal of Bone and Joint Surgery* [3]. The original meeting focused on compiling the basic science evidence supporting sexual dimorphism in musculoskeletal health and developing priorities for future research. Clinical examples were used primarily to illustrate the relevance of this critical topic. This issue of the *Orthopedic Clinics of North America* picks up where the workshop left off, extending the discussion of clinical topics across the broad spectrum of musculoskeletal health. As a prelude and introduction to the issue, a synopsis of the workshop findings follows.

Molecular, cellular, and matrix biology

In 2001, the IOM published a study highlighting two mechanisms that help explain how sex differences in disease presentation arise. First, hormone-responsive genes are influenced by the hormonal milieu of the male and female differently throughout their life spans. Second, genes located on the X and Y sex chromosomes encode proteins that result in variations in biochemical and physiologic pathways. Recent gene expression studies in mice demonstrate that thousands of genes affecting tissues such as liver, fat, muscle, and brain are sexually dimorphic [4].

Cell behavior itself and genes expressed by cells show evidence of sexual dimorphism. Cells derived from females respond differently to estrogen, testosterone, and other steroid hormones than cells derived from males. Differences in response to estrogen and testosterone have been shown for male and female chondrocytes [5], osteoblasts [6], myoblasts [7], colon cells [8], and neuronal cells [9,10].

Dimorphism of the musculoskeletal system is not defined solely by cellular responses; the structure and functions of bone, muscle, tendon, and ligament also exhibit sexual dimorphism. For example, the divergence of skeletal phenotypes between men and women has traditionally been attributed to sex steroid action. New research in mice demonstrates that the inheritance of mechanically relevant bone traits such as strength and toughness also depend on sex, the genotype of the parental strains, and maternal effects [11].

Growth and development

Sexual dimorphism occurs in musculoskeletal size, shape, and strength. It is particularly marked in developing and aging bone. During development, bone enlarges by depositing new tissue on the periosteal surface, whereas resorption occurs on the endosteal surface. For most bones, this expansion is superimposed on cortical drift such that there is apposition and resorption occurring on the periosteal and endosteal surfaces simultaneously [12]. Prepubertal growth is associated with marked periosteal apposition, which is greater than net endocortical resorption. The net effect is that the cortex increases in thickness as long bones increase in length and diameter. After puberty, no further increase in long bone length can occur, but periosteal apposition continues in both sexes, very slowly enlarging the diameter of the long bones. As periosteal apposition is less than net endocortical resorption, the cortical thickness begins to decrease. This cortical thinning occurs to a lesser extent in men than in women [12].

Of particular interest are differences in femoral remodeling. From about the third to the seventh decade, men, but not women, show a fairly uniform increase in subperiosteal area [13]. Furthermore, in the 30- to 70-year age range, osteoclasts produce greater numbers of resorption cavities in women than in men. These cavities in women are also smaller in diameter than those of men [14].

Fractures and bone strength

Bone modeling and remodeling determine bone strength, which determines bone's resistance to fractures. The prevalence of fractures differs

---
[1] Two definitions in the IOM report are germane to this article. Sex is defined as "the classification of living things, generally as male or female according to their reproductive organs and functions assigned by the chromosomal complement" and gender is defined as "a person's self-presentation as male or female, or how that person is responded to by social institutions on the basis of the individual's gender presentation."

between men and women [15]. Falls are a major contributor to fracture [16]; injury severity in both sexes is related to the height of the fall, fall direction, configuration of the body during impact, lack of protective responses, and low bone strength [16–18]. Factors related to the initiation of a fall, however, such as the ability to recover balance, differ between men and women [19–21].

The strength of bones is based on their shape, size, connectivity, and material properties [11]. The genes that control these traits may be different for men and women [22,23]. The pathogenesis of bone fragility in old age is heterogeneous. Reduced bone size in women may be the result of reduced periosteal apposition during growth, aging, or both, and the reduced volumetric bone-mineral density may be the result of reduced peak mineral accrual, age-related bone loss, reduced periosteal apposition, or several of these processes. Men and women have similar peak volumetric bone-mineral densities in young adulthood, but men have larger bones. Although the loads on the vertebral body are greater, the load per unit area (stress) does not differ by sex [24].

Sexual dimorphism in the prevalence of vertebral fractures may be more the result of sex differences in age-related bone gain than of age-related bone loss. Fewer men than women are at risk for fracture, largely because fewer men develop deficits in the structural determinants of bone strength below a level at which loads can exceed the ability of the bone to tolerate them [15,25].

Muscle, tendon, and ligament injury

Women have a higher prevalence of ankle sprains, certain spine injuries, and frozen shoulder than men. Here, too, development likely plays a critical role. For example, before puberty, girls and boys have similar jumping and landing strategies. Boys and girls land similarly, with knees wide apart. After puberty, boys continue to land with knees wide apart, but girls land with the lower extremities in valgus [26].

Several factors have been proposed to explain the increased prevalence of anterior cruciate ligament (ACL) injury in female athletes [27], including sex differences in biomechanical [28], neuromuscular [26], anatomic [29,30], and hormonal [31,32] influences. It seems likely that all of these factors play some role in the increased susceptibility of females to ACL injury.

Sex hormones play a role in determining the size and shape of soft tissues, as evidenced by the changes that occur during puberty. Neuromuscular changes that occur at puberty also reflect the importance of sex hormones in the prevalence of ACL injury.

Musculoskeletal stability describes the capacity of a multisegment limb to maintain and control dynamic alignment under functional load to avoid strain injury to the passive tissues of the joints [33]. Functional stability is maintained by means of muscle recruitment and neural reflex response. The critical components of stabilizing control include the recruitment patterns of voluntary muscles, intrinsic biomechanical impedance (stiffness) of actively contracting muscles, sensory proprioception, and reflex response. Sex differences have been identified in each component.

Reflex response also plays an important role in the dynamic alignment and stability of the musculoskeletal structure and includes components such as sensory proprioception and reflex gain. Ligaments contribute to sensory proprioception. Sex differences in ligament laxity are well established [34] and partially attributed to hormonal differences [35]. They may explain the reported differences in passive tissue laxity and reflex response following prolonged or cyclic joint loading [36].

The basis for the sex differences in ligament laxity is controversial. Sex differences in tissue mechanics may explain reported differences in passive tissue laxity [33] and reflex response [36] following prolonged or cyclic joint loading and fatiguing activity, respectively.

Sexual dimorphism in neurovascular disease and cancer

Stroke

The prevalence of stroke is lower in women than in men until advanced age [37–40]. This neuroprotection is lost within 10 years after menopause, an observation attributed to age-related estrogen loss.

Because of data suggesting that estrogen was protective against vascular disease and stroke, estrogen became one of the commonest medications prescribed in the United States. The results from clinical trials, however, led to questions about the use of estrogen for primary or secondary prevention of vascular disease. It is now recommended that women who have postmenopausal symptoms be treated with hormone replacement therapy at the lowest effective dose for the shortest possible time [41–43].

Despite these new clinical findings, basic research continues to demonstrate that estrogen is neuroprotective in a variety of laboratory and animal models [44]. The understanding of sex differences in cerebral ischemia is a critical starting point for the development of effective neuroprotective strategies.

Cancer

Men experience disproportionately more primary bone tumors than women, and women have more bone metastases than men. Men have 60% more primary musculoskeletal tumors than women. (Giant-cell tumor, surface osteosarcoma, and soft-tissue desmoid tumors are the exceptions to this rule.)

The skeleton is the third most common site of metastatic cancer, and up to one half of all patients who have cancer have metastasis to bone [45]. A change in bone structural properties as a result of tumor-induced osteolysis may determine fracture risk in patients who have skeletal metastases [46]. Metastatic cancer can affect the material properties of bone tissue directly, by stimulating bone resorption (eg, breast cancer) or bone formation (eg, prostate cancer) or by changing bone geometry by remodeling large regions of bone infiltrated by metastatic cancer cells.

Toxicity to chemotherapeutics differs by sex and estrus status [47,48]. The severity of toxicity may also be influenced by sex [49]. Genetic and human leukocyte antigen tissue type and metabolic differences cause a differential response and toxicity [50].

Pain and analgesics

Women are more sensitive to, less tolerant of, and more able to discriminate pain [51] than men. These differences may be due to different neural circuits, transmitters, receptors, and genes that are relevant to pain modulation in men and women.

Differences in the efficacy of analgesics in men and women have also been described [52]. There are sex differences in response to and use of μ-receptor opiates [53,54], κ-acting opiates [55], and anti-inflammatory medications [56] in animals and humans.

Nonsteroidal anti-inflammatory drugs and selective cyclooxygenase-2 inhibitors are among the most commonly prescribed medications [57]. There are no data on whether their use affects fracture healing and bone ingrowth differently in men and women, which may have important clinical implications because young women are encouraged to use these drugs for relief of pain associated with sports injuries and the management of menstrual symptoms.

Nonsteroidal anti-inflammatory agents are widely used in the treatment of osteoarthritis [58], whereas newer immunosuppressive drugs are more commonly used in autoimmune diseases such as rheumatoid arthritis. Many of these autoimmune diseases show a female predilection [59].

An agenda for musculoskeletal health

Workshop participants at the 2004 NIH-AAOS workshop "Influence of Sex and Gender on Musculoskeletal Health" echoed the findings of the Task Force on the NIH Women's Health Research Agenda for the 21st Century [60] and urged multidisciplinary investigations across the life span. The attendees then recast the recommendations developed by the IOM to make them germane to the challenges faced in orthopedics. They identified six priorities for musculoskeletal research [3]:

1. Monitor sex differences and similarities for all musculoskeletal diseases and conditions, including diagnosis and treatment, that affect both sexes.
2. Conduct longitudinal and cross-sectional studies of musculoskeletal diseases and conditions so that results can be analyzed by sex.
3. Make sex-specific data in musculoskeletal diseases and conditions more readily available.
4. Mine cross-species information and develop relevant in vivo and in vitro models that incorporate the biologic clock. Stratify human and animal studies of musculoskeletal diseases and conditions based on sex.
5. Expand research on sex differences in neural organization and function, pain, and analgesia with respect to musculoskeletal diseases and conditions.
6. Promote research on sex differences at the molecular, cellular, and tissue levels, with specific emphasis on musculoskeletal diseases and conditions.

These six recommendations remain valid today, but the orthopedic profession continues to progress, as the reader will see in the articles that follow.

References

[1] Arias E. United States life tables, 2001. Natl Vital Stat Rep 2004;52:1–38.

[2] Wizemann TM, Pardue ML, editors. Institute of Medicine (US) Committee on Understanding the Biology of Sex and Gender Differences. Exploring the biological contributions to human health: does sex matter? Washington, DC: National Academy Press; 2001.

[3] Tosi LL, Boyan BD, Boskey AL. Does sex matter in musculoskeletal health? The influence of sex and gender on musculoskeletal health. J Bone Joint Surg Am 2005;87(7):1631–47.

[4] Yang X, Schadt EE, Wang S, et al. Tissue-specific expression and regulation of sexually dimorphic genes in mice. Genome Res 2006;16(8):995–1004.

[5] Kinney RC, Schwartz Z, Week K, et al. Human articular chondrocytes exhibit sexual dimorphism in their responses to 17beta-estradiol. Osteoarthritis Cartilage 2005;13:330–7.

[6] Ishida Y, Heersche JN. Progesterone stimulates proliferation and differentiation of osteoprogenitor cells in bone cell populations derived from adult female but not from adult male rats. Bone 1997; 20:17–25.

[7] Loukotova J, Kunes J, Zicha J. Cytosolic free calcium response to angiotensin II in aortic VSMC isolated from male and female SHR. Physiol Res 1998; 47:507–10.

[8] Harvey BJ, Higgins M. Nongenomic effects of aldosterone on Ca2+ in M-1 cortical collecting duct cells. Kidney Int 2000;57:1395–403.

[9] Nilsen J, Mor G, Naftolin F. Estrogen-regulated developmental neuronal apoptosis is determined by estrogen receptor subtype and the Fas/Fas ligand system. J Neurobiol 2000;43:64–78.

[10] Ramirez O, Jimenez E. Sexual dimorphism in rat cerebrum and cerebellum: different patterns of catalytically active creatine kinase isoenzymes during postnatal development and aging. Int J Dev Neurosci 2002;20:627–39.

[11] van der Meulen MC, Jepsen KJ, Mikic B. Understanding bone strength: size isn't everything. Bone 2001;29:101–4.

[12] Enlow DH. Principles of bone remodeling; an account of post-natal growth and remodeling processes in long bones and the mandible. Springfield (IL): Thomas; 1963.

[13] Filardi S, Zebaze RM, Duan Y, et al. Femoral neck fragility in women has its structural and biomechanical basis established by periosteal modeling during growth and endocortical remodeling during aging. Osteoporos Int 2004;15:103–7.

[14] Croucher PI, Gilks WR, Compston JE. Evidence for interrupted bone resorption in human iliac cancellous bone. J Bone Miner Res 1995;10:1537–43.

[15] Duan Y, Turner CH, Kim BT, et al. Sexual dimorphism in vertebral fragility is more the result of gender differences in age-related bone gain than bone loss. J Bone Miner Res 2001;16:2267–75.

[16] Hayes WC, Myers ER, Morris JN, et al. Impact near the hip dominates fracture risk in elderly nursing home residents who fall. Calcif Tissue Int 1993;52: 192–8.

[17] Greenspan SL, Myers ER, Maitland LA, et al. Fall severity and bone mineral density as risk factors for hip fracture in ambulatory elderly. JAMA 1994;271:128–33.

[18] Nevitt MC, Cummings SR. Type of fall and risk of hip and wrist fractures: the study of osteoporotic fractures. J Am Geriatr Soc 1993;41:1226–34.

[19] Pavol MJ, Owings TM, Foley KT, et al. The sex and age of older adults influence the outcome of induced trips. J Gerontol A Biol Sci Med Sci 1999;54: M103–8.

[20] Wojcik LA, Thelen DG, Schultz AB, et al. Age and gender differences in single-step recovery from a forward fall. J Gerontol A Biol Sci Med Sci 1999;54: M44–50.

[21] Wojcik LA, Thelen DG, Schultz AB, et al. Age and gender differences in peak lower extremity joint torques and ranges of motion used during single-step balance recovery from a forward fall. J Biomech 2001;34:67–73.

[22] Klein RF, Turner RJ, Skinner LD, et al. Mapping quantitative trait loci that influence femoral cross-sectional area in mice. J Bone Miner Res 2002;17: 1752–60.

[23] Karasik D, Cupples LA, Hannan MT, et al. Age, gender, and body mass effects on quantitative trait loci for bone mineral density: the Framingham Study. Bone 2003;33:308–16.

[24] Seeman E. During aging, men lose less bone than women because they gain more periosteal bone, not because they resorb less endosteal bone. Calcif Tissue Int 2001;69:205–8.

[25] Duan Y, Seeman E, Turner CH. The biomechanical basis of vertebral body fragility in men and women. J Bone Miner Res 2001;16:2276–83.

[26] Ford KR, Myer GD, Hewett TE. Valgus knee motion during landing in high school female and male basketball players. Med Sci Sports Exerc 2003;35: 1745–50.

[27] Arendt E, Dick R. Knee injury patterns among men and women in collegiate basketball and soccer. NCAA data and review of literature. Am J Sports Med 1995;23:694–701.

[28] Nisell R. Mechanics of the knee. A study of joint and muscle load with clinical applications. Acta Orthop Scand Suppl 1985;216:1–42.

[29] Tillman MD, Smith KR, Bauer JA, et al. Differences in three intercondylar notch geometry indices between males and females: a cadaver study. Knee 2002;9:41–6.
[30] Uhorchak JM, Scoville CR, Williams GN, et al. Risk factors associated with noncontact injury of the anterior cruciate ligament: a prospective four-year evaluation of 859 West Point cadets. Am J Sports Med 2003;31:831–42.
[31] Slauterbeck JR, Fuzie SF, Smith MP, et al. The menstrual cycle, sex hormones, and anterior cruciate ligament injury. J Athl Train 2002;37:275–8.
[32] Wojtys EM, Huston LJ, Boynton MD, et al. The effect of the menstrual cycle on anterior cruciate ligament injuries in women as determined by hormone levels. Am J Sports Med 2002;30:182–8.
[33] McGill SM, Cholewicki J. Biomechanical basis for stability: an explanation to enhance clinical utility. J Orthop Sports Phys Ther 2001;31:96–100.
[34] Huston LJ, Wojtys EM. Neuromuscular performance characteristics in elite female athletes. Am J Sports Med 1996;24:427–36.
[35] Slauterbeck JR, Hardy DM. Sex hormones and knee ligament injuries in female athletes. Am J Med Sci 2001;322:196–9.
[36] Moore BD, Drouin J, Gansneder BM, et al. The differential effects of fatigue on reflex response timing and amplitude in males and females. J Electromyogr Kinesiol 2002;12:351–60.
[37] Thorvaldsen P, Kuulasmaa K, Rajakangas AM, et al. Stroke trends in the WHO MONICA project. Stroke 1997;28:500–6.
[38] Sudlow CL, Warlow CP. Comparable studies of the incidence of stroke and its pathological types: results from an international collaboration. International Stroke Incidence Collaboration. Stroke 1997;28:491–9.
[39] Viscoli CM, Brass LM, Kernan WN, et al. A clinical trial of estrogen replacement therapy after ischemic stroke. N Engl J Med 2001;345:1243–9.
[40] Wyller TB. Stroke and gender. J Gend Specif Med 1999;2:41–5.
[41] Wassertheil-Smoller S, Hendrix SL, Limacher M, et al. Effect of estrogen plus progestin on stroke in postmenopausal women: the Women's Health Initiative: a randomized trial. JAMA 2003;289:2673–84.
[42] Anderson GL, Limacher M, Assaf AR, et al. Women's Health Initiative Steering Committee. Effects of conjugated equine estrogen in postmenopausal women with hysterectomy: the Women's Health Initiative randomized controlled trial. JAMA 2004;291:1701–12.
[43] Espeland MA, Rapp SR, Shumaker SA, et al. Women's Health Initiative Memory Study. Conjugated equine estrogens and global cognitive function in postmenopausal women: Women's Health Initiative Memory Study. JAMA 2004;291:2959–68.

[44] Alkayed NJ, Murphy SJ, Traystman RJ, et al. Neuroprotective effects of female gonadal steroids in reproductively senescent female rats. Stroke 2000;31:161–8.
[45] Surveillance, Epidemiology, and End Results (SEER) Program. SEER*Stat Database: US mortality data—all COD, public-use with state, total US (1969–2001), National Cancer Institute, Division of Cancer Control and Population Sciences, Surveillance Research Program, Cancer Statistics Branch. Available at: www.seer.cancer.gov. Released April 2004. Underlying mortality data provided by the National Center for Health Statistics. Available at: www.cdc.gov/nchs. Accessed October 22, 2006.
[46] Hong J, Cabe GD, Tedrow JR, et al. Failure of trabecular bone with simulated lytic defects can be predicted non-invasively by structural analysis. J Orthop Res 2004;22:479–86.
[47] Postma A, Elzenga NJ, Haaksma J, et al. Cardiac status in bone tumor survivors up to nearly 19 years after treatment with doxorubicin: a longitudinal study. Med Pediatr Oncol 2002;39:86–92.
[48] Hequet O, Le QH, Moullet I, et al. Subclinical late cardiomyopathy after doxorubicin therapy for lymphoma in adults. J Clin Oncol 2004;22:1864–71.
[49] Milano G, Ferrero JM, Francois E. Comparative pharmacology of oral fluoropyrimidines: a focus on pharmacokinetics, pharmacodynamics and pharmacomodulation. Br J Cancer 2004;91:613–7.
[50] Macdonald JS. Vive la difference: sex and fluorouracil toxicity. J Clin Oncol 2002;20:1439–41.
[51] Fillingim RB, Ness TJ. Sex-related hormonal influences on pain and analgesic responses. Neurosci Biobehav Rev 2000;24:485–501.
[52] Craft RM. Sex differences in opioid analgesia: "from mouse to man. Clin J Pain 2003;19:175–86.
[53] Cepeda MS, Carr DB. Women experience more pain and require more morphine than men to achieve a similar degree of analgesia. Anesth Analg 2003;97:1464–8.
[54] Sarton E, Olofsen E, Romberg R, et al. Sex differences in morphine analgesia: an experimental study in healthy volunteers. Anesthesiology 2000;93:1245–54 [discussion: 6A].
[55] Gear RW, Gordon NC, Miaskowski C, et al. Sexual dimorphism in very low dose nalbuphine postoperative analgesia. Neurosci Lett 2003;339:1–4.
[56] Walker JS, Carmody JJ. Experimental pain in healthy human subjects: gender differences in nociception and in response to ibuprofen. Anesth Analg 1998;86:1257–62.
[57] Goodman SB. Use of COX-2 specific inhibitors in operative and nonoperative management of patients with arthritis. Orthopedics 2000;23:S765–8.
[58] Gillette JA, Tarricone R. Economic evaluation of osteoarthritis treatment in Europe. Expert Opin Pharmacother 2003;4:327–41.
[59] Lockshin MD. Why do women have rheumatic disease? Scand J Rheumatol Suppl 1998;107:5–9.

[60] United States Department of Health and Human Services, Public Health Service, National Institutes of Health. Agenda for Research on Women's Health for the 21st Century. A report of the Task Force on the NIH Women's Health Research Agenda for the 21st Century. Vol. 1. Bethesda (MD): National Institutes of Health; 1999. NIH publication 99 4385.

Upper Extremity: Emphasis on Frozen Shoulder

Monique A. Sheridan, BA[a], Jo A. Hannafin, MD, PhD[a,b],*

[a]*Women's Sports Medicine Center, Hospital for Special Surgery, 535 East 70th Street, New York, NY 10021, USA*
[b]*Weill Medical College of Cornell University, 1300 York Avenue, New York, NY 10021, USA*

Epidemiology

Primary adhesive capsulitis, or frozen shoulder, is a condition characterized by gradual loss of active and passive glenohumeral motion. The prevalence of frozen shoulder is slightly greater than 2% in the general population [1], affecting persons older than 40 years [2]. Approximately 70% of patients presenting with adhesive capsulitis are women, and 20% to 30% of those affected will develop adhesive capsulitis in the opposite shoulder [1]. Despite the female preponderance of patients who have adhesive capsulitis, the role of sex has not been investigated thoroughly and the etiology remains unknown. The lack of consistency in the published literature also reflects a lack of understanding of causation. As a result, management of frozen shoulder is controversial, and many treatment options, both operative and nonoperative, are available.

Duplay [3] initially identified a stiff shoulder as "periarthritis" in 1872, followed by Codman [4] who labeled the condition "frozen shoulder" in 1934. It was not until 1945, that Neviaser [5] coined the term "adhesive capsulitis," recognizing pathologic changes in the capsule. Disagreement remains in the literature as to whether the underlying pathologic process is an inflammatory condition [5–8] or a fibrosing condition [9]. Progression of the disease has been shown to include contracture of the coracohumeral ligament [10–12]. The terms *frozen shoulder* and *adhesive capsulitis* are used intermittently throughout this article because they are the two most common terms used to describe the condition.

The literature contains limited reports of differences in the anatomy and physiology or the incidence of shoulder disorders between men and women. Shoulder injuries are more common in women due to increased ligament and joint laxity, relatively weaker upper body strength, and shorter long bone length [13]; however, because the etiology of adhesive capsulitis of the shoulder is unknown, it is unclear why the condition occurs more frequently in women. Frozen shoulder has also been documented to be more common and more difficult to treat effectively in patients who have diabetes [14–17], thyroid disease [18,19], and autoimmune disease [20,21], and a statistically significant association between frozen shoulder and Dupuytren's disease has been found [22].

Richards and colleagues [23] examined the relationship between adhesive capsulitis and acromial morphology. Of the 69 shoulders with adhesive capsulitis, 75.4% carried a type II acromion. Acromial morphology, however, was not found to have a statistically significant relationship with adhesive capsulitis because 74.1% of the control group also exhibited a type II acromion. Richards and colleagues [23] concluded that narrowing of the subacromial space caused by the anterior acromial shape was not a cause of primary frozen shoulder. Sex differences were not assessed in this study.

Anatomy and physiology

Adhesive capsulitis can be classified as primary, which is characterized by idiopathic, progressive, painful loss of active and passive shoulder motion; as secondary, which has a similar presentation and progression but results from a known intrinsic or extrinsic cause; or as

* Corresponding author.
E-mail address: HannafinJ@hss.edu (J.A. Hannafin).

secondary shoulder stiffness following surgical intervention. Intrinsic factors implicated in the causation of secondary adhesive capsulitis include rotator cuff tears, bursitis, and tendonitis, whereas extrinsic factors generally relate to trauma.

Neviaser and Neviaser [24,25] described four stages of adhesive capsulitis. Hannafin and colleagues [9] described a correlation among the arthroscopic stages described by Nevasier, the clinical examination, and the histologic appearance of capsular biopsy specimens in patients who had stages 1, 2, and 3 adhesive capsulitis. It is imperative to note that these stages represent a continuum of disease rather than discrete, well-defined stages. To date, there are no published data showing anatomic or physiologic differences between men and women in the four stages of adhesive capsulitis.

In stage 1, patients complain of pain with active and passive range of motion. The pain is described as achy at rest and sharp with motion, with symptoms present for less than 3 months. Patients describe night pain and rest pain. There is a progressive loss of motion, with forward flexion, abduction, internal rotation, and external rotation becoming most limited. Upon examination under anesthesia or following intra-articular injection of local anesthetic, there is a significant improvement in range of motion to normal or to minimal loss. Arthroscopic examination reveals a diffuse hypervascular glenohumeral synovitis, often most pronounced in the anterosuperior capsule (Fig. 1). Pathology specimens show rare inflammatory cell infiltrates; a hypertrophic, hypervascular synovitis; and a normal underlying capsule (Fig. 2).

In stage 2, also known as the "freezing stage," symptoms have been present for 3 to 9 months with chronic pain and progressive loss of range of motion. There is still rest pain and night pain, and

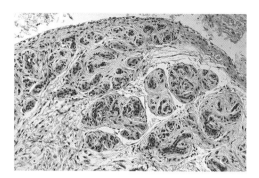

Fig. 2. Hypertrophic, hypervascular synovitis with perivascular and subsynovial scar formation in stage 2 adhesive capsulitis.

significant sleep disturbances may exist. There is significant limitation of forward flexion, abduction, internal rotation, and external rotation. Examination after intra-articular injection of local anesthetic or scalene block reveals relief of pain with partial improvement in range of motion. The motion loss in stage 2 reflects a loss of capsular volume and a response to painful synovitis. Arthroscopic examination reveals a diffuse, pedunculated synovitis and a tight capsule with a rubbery or dense feel on the insertion of the arthroscope (Fig. 3). There is a hypertrophic, hypervascular synovitis with perivascular and subsynovial scar formation and capsular fibroplasia (Fig. 4). No inflammatory infiltrates have been reported in stage 2.

In the "frozen stage" (stage 3), patients experience minimal pain at night or rest (except at the end range of motion) but have significant shoulder stiffness. There is a marked loss of range of motion with a rigid "end feel" on capsular stress. Symptoms have been present for approximately 9 to 15 months. Range of motion remains unchanged when the patient is injected with local

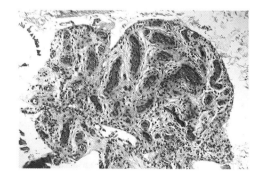

Fig. 1. Biopsy sample from stage 1 adhesive capsulitis demonstrates hypervascular hypertrophic synovium.

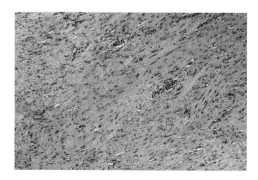

Fig. 3. Capsular fibroplasias seen in deep capsular biopsy tissue from late stage 2 adhesive capsulitis.

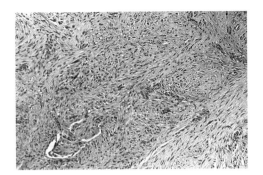

Fig. 4. Capsular biopsy sample demonstrates dense, hypercellular collagenous tissue in stage 3 adhesive capsulitis.

anesthetic or examined under anesthesia, secondary to a profound loss of capsular volume and fibrosis of the glenohumeral joint capsule. Arthroscopic examination reveals remnants of fibrotic synovium that is not hypervascular. Capsular biopsy samples reveal a dense, hypercellular collagenous tissue and a thin synovial layer without significant hypertrophy or hypervascularity.

Stage 4 is known as the "thawing phase" of adhesive capsulitis. During this stage, there is minimal pain and progressive improvement in range of motion resulting from capsular remodeling. Because these patients rarely undergo surgery, there are no arthroscopic or histologic data available for patients who have stage 4 adhesive capsulitis. Determination of the stage of adhesive capsulitis at the time of patient examination is critical and directs the treatment options.

Diagnosis

The diagnosis of idiopathic frozen shoulder is made from history and physical examination when other causes of pain and motion loss are eliminated. It is important to determine from the patient's history the current stage of the condition to determine the appropriate treatment. The physical examination should include an evaluation of the cervical spine, trunk, and both shoulders. Patients presenting with stages 1 and 2 adhesive capsulitis have pain on palpation of the anterior and posterior capsule and describe pain radiating to the deltoid insertion. Night and rest pain are common in the early stages. Sufficient evaluation of active and passive range of motion is necessary to determine the stage of disease and the subsequent efficacy of the treatment.

Glenohumeral motion should be measured while stabilizing the scapula to evaluate glenohumeral versus scapulothoracic range of motion. External rotation, abduction, and internal rotation are most affected. Active range of motion should be measured and recorded with the patient standing. Passive glenohumeral motion is measured with the patient supine and scapulothoracic motion constrained by manual pressure on the acromion.

Routine radiographs are obtained, including anteroposterior views in internal and external rotation, axillary views, and outlet views, to identify other potential causes of a stiff shoulder such as glenohumeral arthritis, calcific tendonitis, or rotator cuff disease. Plain radiographs are generally normal in patients who have frozen shoulder, although there may be evidence of osteopenia. MRI may be useful in recognizing partial or complete rotator cuff tears in patients complaining of shoulder stiffness and pain but it is not routinely needed for the diagnosis of adhesive capsulitis.

Nonoperative treatment

The treatment given to patients presenting with adhesive capsulitis depends on the stage of the disease and clinical symptoms. There are certain basic principles that apply to all stages. Oral nonsteroidal anti-inflammatory medications can be initiated in most patients who present with painful limited range of motion, with other analgesics supplemented as necessary. A combined intra-articular injection of corticosteroid and local anesthetic is also helpful in managing adhesive capsulitis by reducing pain and allowing for increased shoulder movement.

Exercise and physical therapy are often highly recommended to maintain and regain range of motion. Physical therapy, stretching, and other rehabilitation programs are most effective in patients presenting with stage 2 or higher adhesive capsulitis because stage 1 patients often find physical therapy difficult due to inflammation and pain. Therefore, the goal of physical therapy for stage 1 adhesive capsulitis is to interrupt the cycle of inflammation and pain by focusing on modalities for pain, patient education for positioning, and activity modification to balance activity and rest. Stage 2 treatments concentrate on decreasing inflammation and pain and minimizing capsular restriction to minimize loss of motion. The goal is to stretch the capsule sufficiently to allow restoration of normal glenohumeral biomechanics. Physical therapy for patients who have stage 3 adhesive capsulitis is designed to treat significant loss of motion by increasing range of motion through aggressive stretching. Heat promotes muscle relaxation; hydrotherapy may also be used to reduce discomfort after stretching. Most patients have significant improvement by 12 to 16 weeks. Some patients do not improve and may get worse, which may indicate a need for surgical intervention or manipulation.

Physical therapy has proved to be an effective method of treatment for adhesive capsulitis. A recent placebo-controlled study [26] found that a stretching-exercise program in patients who had stage 2 idiopathic adhesive capsulitis successfully reduced pain at rest (84% of subjects) and with activity (73% of subjects). Vermeulen and colleagues [27] later examined the effectiveness of high-grade mobilization techniques (movements into the stiffness zone) and low-grade mobilization techniques (movements in the pain-free zone) in 100 patients who had adhesive capsulitis symptoms for greater than 3 months and at least a 50% decrease in shoulder joint mobility. The study found that both groups improved range of motion within 12 months. The results for the group that underwent high-grade mobilization techniques were superior to those of the low-grade group, but the overall difference between the two groups was small. The study suggests that physical therapy is valuable regardless of intensity due to increased movement and range of motion and decreased stiffness of the shoulder.

In contrast, Diercks and Stevens [28] compared supervised neglect with intensive physical therapy as treatment for patients presenting with adhesive capsulitis symptoms for greater than 3 months and with more than 50% motion restriction of the glenohumeral joint. After 2 years from the start of treatment, 89% of patients treated with supervised neglect had normal or near-normal shoulder function compared with only 63% of subjects treated successfully with intensive physical therapy. In this study, supervised neglect included pendulum and active exercises in the pain-free range and instructions to resume all tolerable activities. The investigators suggested that the intensive physical therapy program had an adverse effect on the natural course of the active disease process. It is important to note that both treatments were more than 50% effective and that there was no long-term evidence of efficacy of either method.

Although there are no published data that examine the role of sex in response to treatment, Griggs and colleagues [26] incorporated a mental health survey into a study evaluating the efficacy of a physical therapy program for patients who had stage 2 adhesive capsulitis. Sixty-six patients completed a Short Form-36 Health Survey to compare the subjects' mental health with the established population norms. In subjects whose pain decreased over the 12-month stretching-exercise period, the Health Survey demonstrated that male subjects' scores were not significantly different from those in the general male population. In contrast, female subjects scored higher in vitality and mental health compared with the general female population. The investigators suggested that female patients who overcome adhesive capsulitis do not have an intrinsic emotional, psychologic, or personality disorder. The study also suggests that the women were satisfied with the results of the stretching-exercise program, resulting in a positive overall mental health status. It remains unclear whether women find the symptoms of acute adhesive capsulitis more difficult to manage. These data also raise the question whether the condition is more severe in women and whether the improvements in pain and range of motion after treatment are more marked in women.

Intra-articular injections of steroid and local anesthetics are extremely useful tools in the diagnosis and treatment of adhesive capsulitis. An injection not only provides pain relief but also restores motion and function of the shoulder, which helps to reduce the level of stiffness. There is extensive information regarding the efficacy of intra-articular corticosteroids. Bulgen and colleagues [29] randomized patients to treatment with corticosteroids, physical therapy, or benign neglect. The initial response to treatment was most marked in patients treated with corticosteroids. Van der Windt and coworkers [30] confirmed that corticosteroids were superior to physiotherapy in providing faster relief of symptoms. At 7 weeks, 77% of patients who received corticosteroid injections were considered treatment successes versus only 46% of patients receiving physiotherapy. Other studies [31,32], however, show that there is no significant difference in benefits of corticosteroid injections or physiotherapy treatments. In these trials, however, the stage of adhesive capsulitis treated is not known. In most trials, there was no long-term difference (2-year follow-up) between treatment groups, as might be expected in a self-limiting condition. Therefore, corticosteroid injections are probably effective only in the short-term course of managing frozen shoulder syndrome. Recently, a randomized, placebo-controlled study [33] determined that corticosteroids provide significant short-term benefits in adhesive capsulitis but the benefits do not extend beyond 6 months of treatment. Depending on the duration of symptoms and the stage of the condition, patients may require additional treatments or surgical referral.

Recent research has also shown that when corticosteroids are combined with other treatment methods such as physiotherapy, the results are superior. Carette and colleagues [34] randomized 93 subjects to four treatment groups: corticosteroid injection combined with physiotherapy, corticosteroid injection alone, physiotherapy alone, and placebo. By the end of the study, all four groups had reached a comparable level of improvement in pain and range of motion, which is consistent with the natural course of the disease. The combined corticosteroid injection and physiotherapy treatment, however, proved to be most effective in providing quick improvements in shoulder range of motion compared with either treatment alone. Ryans and colleagues [35] supported that treatment success is more likely in patients who have frozen shoulder receiving a combination of physiotherapy and corticosteroid injections. The study noted that although corticosteroid injections significantly improve shoulder-related disability at 6 weeks, a physiotherapy program effectively improves the range of external motion at 6 weeks of treatment. When selecting a treatment method for adhesive capsulitis of the shoulder, it is extremely important to consider the patient's symptoms and stage of the condition because each patient's treatment should be individualized and tailored appropriately to patient needs. Patients who have stage 1 or 2 adhesive capsulitis should receive intra-articular steroid injections based on our understanding of the inflammatory component of those stages. Stage 3 would not be expected to respond to intra-articular steroid; thus, treatment with this modality is not indicated. The importance of stage of disease in treatment may reflect the mixed results in the aforementioned trials.

The role of sex in the response to nonoperative treatment of frozen shoulder syndrome has yet to be explored. In the many trials evaluating conservative treatments for adhesive capsulitis, such as physical therapy and corticosteroid injections, studies report a majority of female subjects, as seen in Table 1 [26–28,30,32–35]. These data are consistent with Binder and colleagues' [1] report that adhesive capsulitis of the shoulder occurs more frequently in women than in men.

Although van der Windt and colleagues' trial [30] did not assess outcome of treatment with regard to sex, the investigators observed sex-specific reactions to the corticosteroid treatments. Of the

Table 1
Nonoperative treatment of adhesive capsulitis of the shoulder

Study	Total number of subjects presenting with adhesive capsulitis of the shoulder	Female subjects presenting with adhesive capsulitis of the shoulder n (%)	Trial treatment methods
Ryans et al, 2005 [35]	80	46 (57.5)	Physiotherapy, corticosteroid injections
Carette et al, 2003 [34]	93	55 (59.1)	Physiotherapy, corticosteroid injections
Buchbinder et al, 2004 [33]	49	35 (71.4)	Corticosteroid injections
Arslan & Çeliker, 2001 [32]	20	10 (50.0)	Physical therapy, corticosteroid injections
Dierks & Stevens, 2004 [28]	77	40 (51.9)	Physical therapy, supervised neglect
Van der Windt et al, 1998 [30]	109	58 (53.2)	Physiotherapy, corticosteroid injections
Vermeulen et al, 2006 [27]	100	66 (66.0)	Physical therapy
Griggs et al, 2000 [26]	75	58 (75.3)	Physical therapy

58 female subjects, several women experienced adverse reactions to corticosteroids. Nine women reported facial flushing and 6 reported menstrual bleeding, 2 of whom were postmenopausal. Doctors and patients must be aware of such adverse reactions when administering intra-articular steroid injections to female patients presenting with adhesive capsulitis of the shoulder.

Operative treatment

Operative treatment is recommended for patients who do not respond to or who demonstrate little improvement after conservative treatment. Treatment options include closed manipulation under anesthesia, arthroscopic capsular release, and open surgical capsular release.

Closed manipulation under anesthesia

Closed manipulation serves as an effective operative treatment method for most patients presenting with frozen shoulder, as demonstrated in the literature [36–39]. There is no published literature that documents the role of sex in closed manipulation. Of the research available, however, it is notable that the number of female subjects is significantly greater than that of male subjects, as demonstrated in Table 2. This analysis supports the common theory that adhesive capsulitis is more common in women than in men. A high incidence in failure with closed manipulation treatment has been reported in patients who have diabetes [39]. Contraindications also include significant osteopenia, recent surgical repair of soft tissues about the shoulder, or the presence of fractures or neurologic injury.

Closed manipulation is performed after placement of scalene block anesthesia or induction of general anesthesia. The scapula is stabilized with one hand while the humerus is grasped just above the elbow with the other hand. Initially, the adducted shoulder is externally rotated and then abducted in the coronal plane. Next, the shoulder is externally rotated in abduction and then internally rotated in abduction. The shoulder then is forward flexed and finally brought back into adduction and internally rotated. Frequently, there is palpable and audible yielding of the soft tissue as motion is restored in the different planes.

The authors do not advocate primary closed manipulation of the shoulder as the treatment of

Table 2
Operative treatment of adhesive capsulitis of the shoulder

Study	Total number of subjects presenting with adhesive capsulitis of the shoulder	Female subjects presenting with adhesive capsulitis of the shoulder n (%)	Trial treatment methods
Farrell et al, 2005 [37]	25	16 (64.0)	Manipulation under anesthesia
Othman & Taylor, 2002 [36]	64	37 (57.8)	Manipulation under anesthesia
Hamdan & Al-Essa, 2003 [39]	100	61 (61.0)	Manipulation under anesthesia with or without saline injection or steroid injection
Dodenhoff et al, 2000 [38]	37	27 (73.0)	Manipulation under anesthesia
Anderson et al, 1998 [40]	24	13 (54.2)	Manipulation under anesthesia, arthroscopic inspection
Watson et al, 2000 [42]	73	42 (57.5)	Arthroscopic capsular release
Ide & Takgi, 2004 [45]	42	24 (57.1)	Arthroscopic capsular release
Nicholson, 2003 [44]	68	41 (60.3)	Arthroscopic capsular release
Klinger et al, 2002 [41]	36	22 (61.1)	Arthroscopic capsular release
Berghs et al, 2004 [43]	25	13 (52.0)	Arthroscopic capsular release

choice for patients who have adhesive capsulitis but prefer an arthroscopic inspection before manipulative treatment is performed. An earlier Dutch study [40] supports the authors' position, demonstrating that manipulation coupled with arthroscopy is effective in reducing stiffness and pain resulting from adhesive capsulitis. In this study, all patients underwent arthroscopy under general anesthesia with passive motion determined at onset. Initial findings at arthroscopy were a reduced intra-articular volume and diffuse synovitis. The arthroscopy after manipulation revealed that 79% of cases had a rupture of the capsule adjacent to the anterior inferior rim. Physical therapy exercises were included for 2 to 6 weeks following surgery. At 12-months' follow-up, 75% of patients had satisfactory results, with no significant differences in outcome between patients who had primary and secondary adhesive capsulitis, including those who had diabetes.

Arthroscopic capsular release

Arthroscopy is becoming increasingly more popular in the operative treatment of adhesive capsulitis. Arthroscopic intervention permits evaluation of glenohumeral or subacromial disease, synovectomy in stage 2 adhesive capsulitis, and facilitates a precise capsular release in stage 3 adhesive capsulitis. This approach is efficacious in improving shoulder range of motion and function and in reducing recovery time from frozen shoulder syndrome [41–45]. As with other treatment methods, sex differences as related to arthroscopic outcome have not been examined in the literature. As expected, however, there are more female subjects in these studies than male subjects, as detailed in Table 2. Therefore, the same question arises as with treatment by closed manipulation: Do more women than men exhibit persistent stiffness and pain resulting in operative treatment? To answer this question, a prospective cohort study for all patients presenting with a diagnosis of adhesive capsulitis is needed. As arthroscopy gains popularity, it is also important to examine whether women respond differently to arthroscopic treatment than men and whether recovery periods vary between sexes. Gerber and colleagues [46] addressed the tendency of patients to achieve excellent results as related to sex. In this study, the male-to-female ratio was unusual, with 37 men and 8 women. Patients were selected from a group of patients who had adhesive capsulitis and who had not responded favorably to conservative treatment at a shoulder clinic in Switzerland. The study provides no explanation for or insight into the unusual ratio of men to women, but the investigators cited that sex was not a significant predictor of arthroscopic outcome.

Open surgical capsular release

Patients who have adhesive capsulitis and who have failed to respond to closed manipulation or arthroscopic release are sometimes recommended an open surgical release to improve the condition. This treatment method, however, is rare among surgeons in the United States. Although an open surgery allows greater access to the humeroscapular region and facilitates palpation, the disadvantages include postsurgical stiffness, lengthened postoperative recovery time, decreased pain control, and restrictions on physical therapy and rehabilitation exercises. There are limited published data on the efficacy of open surgical releases in the treatment of adhesive capsulitis. Omari and Bunker [12] and Ozaki and colleagues [10] reported excellent results in patients who failed to improve with conservative treatment or closed manipulation; however, each study included limited numbers (25 and 17 subjects, respectively), and no sex analysis was conducted. Although open surgical release is effective in treating adhesive capsulitis, arthroscopy remains the treatment method of choice due to its precision and reduced postoperative recovery time.

Summary

Continued research is indicated to fully understand the etiology of adhesive capsulitis of the shoulder and the response to treatment. There are currently limited published data differentiating the progression and outcomes of treatment of frozen shoulder syndrome, and limited data examining sex differences in adhesive capsulitis. We need to determine why women are at increased risk for development of adhesive capsulitis. Do fluctuations in hormone levels in the premenopausal years somehow trigger this syndrome, or alternatively, do they determine the individual response to the biologic trigger? Is there an autoimmune component to this disease? If so, scientific research may be directed toward determining the role of hormone replacement therapy in the prevention or development of adhesive capsulitis. Do men and women respond comparably to conservative and operative treatment? Prospective clinical trials are needed to determine the efficacy of these treatment regimens for men and women to refine the scientific methods of diagnosis and treatment appropriate for each sex.

References

[1] Binder A, Bulgen DY, Hazelman BL, et al. Frozen shoulder: a long-term prospective study. Ann Rheum Dis 1986;43:288–92.

[2] Lloyd-Roberts G, French P. Periarthritis of the shoulder: a study of the disease and its treatment. BMJ 1959;1:1569–74.

[3] Duplay S. De la peri-arthrite scapulo-humerale et des raideurs de l'épaule qui en sont la consequence [What is the outcome of scapulohumeral periarthritis and stiffness of the shoulder]. Arch Gen Med 1872;20:513–4 [in French].

[4] Codman E. Rupture of the supraspinatus tendon and other lesions in or about the subacromial bursa. In: The shoulder. Boston: Thomas Todd; 1934.

[5] Neviaser J. Adhesive capsulitis of the shoulder: a study of the pathological findings in periarthritis of the shoulder. J Bone Joint Surg 1945;27:211–22.

[6] Hannafin JA, DiCarlo ER, Wickiewicz TL, et al. Adhesive capsulitis: capsular fibroplasias of the glenohumeral joint. J Shoulder Elbow Surg 1994; 3(Suppl):5 [abstract].

[7] Rodeo S, Hannafin J, Tom J, et al. Immunolocalization of cytokines and their receptors in adhesive capsulitis of the shoulder. J Orthop Res 1997;15:427–36.

[8] Wiley A. Arthroscopic appearance of frozen shoulder. Arthroscopy 1991;7:138–43.

[9] Bunker T, Anthony P. The pathology of frozen shoulder: a Dupuytren-like disease. J Bone Joint Surg 1995;77:677–83.

[10] Ozaki J, Nakagawa Y, Sakurai G, et al. Recalcitrant chronic adhesive capsulitis of the shoulder. J Bone Joint Surg Am 1989;71:1511–5.

[11] Neer CS II, Satterlee C, Dalsey R, Flatow E. The anatomy and potential effects of contracture of the coracohumeral ligament. Clin Orthop 1992;280: 182–5.

[12] Omari A, Bunker TD. Open surgical release for frozen shoulder: surgical findings and results of the release. J Shoulder Elbow Surg 2001;10:353–7.

[13] Holschen JC. The female athlete. South Med Assoc 2004;97:852–8.

[14] Bridgman J. Periarthritis of the shoulder and diabetes mellitus. Ann Rheum Dis 1972;31:69–71.

[15] Miller M, Rockwood C Jr. Thawing the frozen shoulder: The "patient" patient. Orthopedics 1997; 18:849–53.

[16] Ogilvie-Harris D, Myerthall S. The diabetic frozen shoulder: arthroscopic release. Arthroscopy 1997; 13:1–8.

[17] Massoud SN, Pearse EO, Levy O, Copeland SA. Operative management of the frozen shoulder in patients with diabetes. J Shoulder Elbow Surg 2002;11:609–13.

[18] Bowman C, Jeffcoate W, Patrick M. Bilateral adhesive capsulitis, oligarthritis and proximal myopathy as presentation of hypothyroidism. Br J Rheumatol 1998;27:62–4.

[19] Wohlgethan J. Frozen shoulder in hyperthyroidism. Arthritis Rheum 1987;30:936–9.

[20] Bulgen D, Binder A, Hazelman B. Immunological studies in frozen shoulder. J Rheumatol 1982;9: 893–8.

[21] Rizk R, Pinals R. Histocompatibility type and racial incidence in frozen shoulder. Arch Phys Med Rehab 1984;65:33–4.

[22] Smith SP, Devaraj VS, Bunker TD. The associate between frozen shoulder and Dupuytren's disease. J Shoulder Elbow Surg 2001;10:149–51.

[23] Richards DP, Glogau AI, Schwartz M, Harn J. Relation between adhesive capsulitis and acromial morphology. Arthroscopy 2004;20:614–9.

[24] Neviaser RJ. Painful conditions affecting the shoulder. Clin Orthop 1983;173:63–9.

[25] Neviaser RJ, Neviaser TJ. The frozen shoulder. Diagnosis and management. Clin Orthop 1987;223: 59–64.

[26] Griggs SM, Ahn A, Green A. Idiopathic adhesive capsulitis: a prospective functional outcome study of nonoperative treatment. J Bone Joint Surg Am 2000;82:1398–407.

[27] Vermeulen HM, Rozing PM, Obermann WR, et al. Comparison of high-grade and low-grade mobilization techniques in the management of adhesive capsulitis of the shoulder: a randomized controlled trial. Phys Ther 2006;86:355–68.

[28] Diercks RL, Stevens M. Gentle thawing of the frozen shoulder: a prospective study of supervised neglect versus intensive physical therapy in seventy-seven patients with frozen shoulder syndrome followed up for two years. J Shoulder Elbow Surg 2004;13: 499–502.

[29] Bulgen D, Binder A, Hazelman B, et al. Frozen shoulder: prospective clinical study with an evaluation of three treatment regimens. Ann Rheum Dis 1985;43:353–60.

[30] Van der Windt BAWM, Koes BW, Devillé W, et al. Effectiveness of corticosteroid injections versus physiotherapy for treatment of painful stiff shoulder in primary care: randomized trial. BMJ 1998;317: 1292–6.

[31] Dacre JE, Beeney N, Scott DL. Injections and physiotherapy for the painful stiff shoulder. Ann Rheum Dis 1989;48:322–5.

[32] Arslan S, Çeliker R. Comparison of the efficacy of local corticosteroid injection and physical therapy for the treatment of adhesive capsulitis. Rheumatol Int 2001;21:20–3.

[33] Buchbinder R, Hoving JL, Green S, et al. Short course prednisolone for adhesive capsulitis (frozen shoulder or stiff painful shoulder): a randomized,

double blind, placebo controlled trial. Ann Rheum Dis 2004;63:1460–9.
[34] Carette S, Moffet H, Tardif J, et al. Intra-articular corticosteroids, supervised physiotherapy, or a combination of the two in the treatment of adhesive capsulitis of the shoulder: a placebo-controlled trial. Arth Rheum 2003;48:829–38.
[35] Ryans I, Montgomery A, Galway R, et al. A randomized controlled trial of intra-articular triamcinolone and/or physiotherapy in shoulder capsulitis. Rheum 2005;44:529–35.
[36] Othman A, Taylor G. Manipulation under anaesthesia for frozen shoulder. Int Orthop 2002;26:268–70.
[37] Farrell CM, Sperling JW, Cofield RH. Manipulation for frozen shoulder: long-term results. J Shoulder Elbow Surg 2005;14:480–4.
[38] Dodenhoff R, Levy O, Wilson A, et al. Manipulation under anesthesia for primary frozen shoulder: effect on early recovery and return to activity. J Shoulder Elbow Surg 2000;9:23–6.
[39] Hamdan TA, Al-Essa KA. Manipulation under anesthesia for the treatment of frozen shoulder. Int Orthop 2003;27:107–9.
[40] Anderson NH, Søjbjerg JO, Johannsen HV, et al. Frozen shoulder: arthroscopy and manipulation under general anesthesia and early passive motion. J Shoulder Elbow Surg 1998;7:218–22.
[41] Klinger HM, Otte S, Baums M, et al. Early arthroscopic release in refractory shoulder stiffness. Arch Ortho Trauma Surg 2002;122:200–3.
[42] Watson L, Dalziel R, Story I. Frozen shoulder: a 12-month clinical outcome trial. J Shoulder Elbow Surg 2000;9:16–22.
[43] Berghs BM, Sole-Molins X, Bunker TD. Arthroscopic release of adhesive capsulitis. J Shoulder Elbow Surg 2004;13:180–5.
[44] Nicholson GP. Arthroscopic capsular release for stiff shoulders: effect on etiology of outcomes. Arthroscopy 2003;19:40–9.
[45] Ide J, Takgi K. Early and long-term results of arthroscopic treatment for shoulder stiffness. J Shoulder Elbow Surg 2004;13:174–9.
[46] Gerber C, Espinosa N, Perren TG. Arthroscopic treatment of shoulder stiffness. Clin Orthop Rel Res 2001;390:119–28.

How Men and Women Are Affected by Osteoarthritis of the Hand

E. Anne Ouellette, MD, MBA*, Anna-Lena Makowski, HTL

Department of Orthopaedics, D-27, University of Miami Miller School of Medicine, PO BOX 016960, Miami, FL 33101, USA

Osteoarthritis of the hand (HOA) is a major cause of impairment in performing activities of daily living. Symptomatic HOA often affects multiple joints. The distal interphalangeal (DIP) joint is most frequently involved; the carpometacarpal (CMC) joint of the thumb ranks second [1].

Sexual dimorphism in HOA has been linked to increasing age, location of onset, health (ie, weight), response of the articular cartilage to estrogen, topography of the joints, and the presence of sex-specific genetic polymorphisms associated with OA. Currently, the diagnosis and treatment of HOA are essentially the same for men and women. Research on possible sex-specific responses to treatment in HOA has yet to be undertaken. Because of the multifactorial nature of OA etiology, research is needed in many disciplines to further understand the connection between OA in general and its apparent disproportional effect on women.

Epidemiology

Thirty percent of all joints affected by OA are the joints of the hand [2]. The DIP joint demonstrates the highest OA prevalence overall. OA prevalence rates specific for second DIP, third proximal phalangeal (PIP), and first CMC joints have been reported at 35%, 18%, and 21%, respectively [3]. Women have a higher prevalence of symptomatic HOA than do men [1].

In population and radiologic studies, OA in the DIP, PIP, metacarpal phalangeal (MCP), and CMC joints is consistently more prevalent in women than men. There are exceptions to this finding; for example, OA of the wrists and the second, index, and DIP joint is more common in men. The severity of HOA increases with age in women, particularly after age 55 [3–5].

In the Framingham study, the strongest predictor for a symptomatic joint in a hand was the presence of OA in the contralateral joint. This correlation was consistent for men and women. Contralateral joint involvement symmetry occurred more often in women than men, however. For men, the prevalence of symptomatic OA in specific hand joints increased by 14 times if the same joint in the opposite hand also was affected. In women, this number was 29.8 times [1]. Occurrence of OA in subsequent rows and rays of the same hand's joints also has been documented. Disease in a joint in a particular row or ray greatly increases the risk for OA in the next joint in the same row for men and women. DIP OA is strongly associated with the development of PIP OA in the same ray. Additional studies confirm the higher incidence of OA in the DIP, PIP, and CMC-1 joints for women than for men. The MCP joints do not usually have a higher incidence of OA in women, as is the case in men who do have a higher incidence of MCP HOA. There was a correlation with knee OA (62.5%) and hypertension in women with CMC-1 OA. When investigated independently, female sex was associated with CMC-1 OA but not with IP OA [6–8].

HOA is associated with pain and disability more commonly in women than men. When

* Corresponding author.
E-mail address: e.ouellet@med.miami.edu (E.A. Ouellette).

associations between hand pain and disability/potential determinants were examined, OA was a significant cause of disability (Tables 1 and 2). Leisure and personal activities are more affected by the disability caused by HOA with increasing age [9–11].

In summary, HOA occurs more often in women than men. The contralateral side is heavily correlated to HOA in the affected site. HOA causes limitations in function in men and women. Hand disability is also more commonly reported in women than men, but HOA is not necessarily a primary reason for the disability.

Anatomy and physiology

Differences between men and women in the anatomy of the hand include the size and contact areas in individual joints. There are also significant differences in surface curvature, cartilage thickness, and joint congruity in the CMC-1 and metacarpal joints. The maximum contact area for

Table 1
Frequency and univariate analysis of hand pain for selected determinants

Determinant	Frequency (%)	Odds ratio (95% CI)
Rheumatoid arthritis	3.6	12.4 (9.5–16.2)
Pain in the neck/shoulder	22.0	3.5 (3.1–4.0)
Osteoarthritis in any joint	24.2	3.2 (2.8–3.6)
Female sex	61.1	2.6 (2.2–3.0)
Thyroid disease[a]	16.9	1.7 (1.5–2.1)
BMI ≥ 30 kg/m^2	14.5	1.4 (1.2–1.7)
History of fracture (hand/wrist)	13/8	1.4 (1.1–1.7)
Diabetes	6.7	1.3 (1.0–1.6)
Manual occupation	28.3	1.1 (1.0–1.3)
Gout	0.8	1.6 (0.9–2.9)
Stroke	4.6	1.0 (0.7–1.3)
Age >70 years	48.2	1.0 (0.9–1.2)
Parkinson's disease	1.0	1.1 (0.6–2.1)

$N = 7883$; age ≥ 55 years. Home interviews were used to determine hand pain during the past month and the presence of any of the selected determinants.

Abbreviations: BMI, body mass index; CI, confidence interval.

[a] Thyroid disease = hypothyroidism, hyperthyroidism, or other thyroid diseases.

(*Data from* Dahaghin S, Bierma-Zeinstra S, Ginai A, et al. Prevalence and determinants of one month hand pain and hand related disability in the elderly (Rotterdam study). Ann Rheum Dis 2005;64:99–104.)

Table 2
Frequency and univariate analysis of hand disability for selected determinants

Determinant	Frequency (%)	Odds ratio (95% CI)
Age >70 years	48.2	6.4 (5.4–7.6)
Rheumatoid arthritis	3.6	6.3 (4.9–8.1)
Stroke	4.6	5.2 (4.1–6.5)
Female sex	61.1	2.8 (2.4–3.3)
Hand pain	16.8	2.6 (2.3–3.1)
Parkinson's disease	1.0	18.4 (10.9–30.8)
Pain in the neck/shoulder	22.0	2.2 (1.9–2.5)
Manual occupation	28.3	2.0 (1.8–2.3)
Thyroid disease*	16.9	2.0 (1.7–2.3)
Diabetes	6.7	2.4 (2.0–3.0)
History of fracture (hand/wrist)	13.8	1.8 (1.5–2.1)
Osteoarthritis in any joint	24.2	1.6 (1.4–1.9)
BMI ≥ 30 kg/m^2	14.5	1.3 (1.0–1.5)
Gout	0.8	0.9 (0.4–2.0)

$N = 7883$; age ≥ 55 years. Home interviews were used to determine hand disability based on the mean score of eight questions related to function. The presence of any of the selected determinants was also recorded.

Abbreviations: BMI, body mass index; CI, confidence interval.

* Thyroid disease = hypothyroidism, hyperthyroidism, or other thyroid diseases.

(*Data from* Dahaghin S, Bierma-Zeinstra S, Ginai A, et al. Prevalence and determinants of one month hand pain and hand related disability in the elderly (Rotterdam study). Ann Rheum Dis 2005;64:99–104.)

women is less than for men. The trapezial surface is less concave in women. The reduced congruency and thinner cartilage in women may account for the greater prevalence in HOA in women. Because the contact areas are smaller and the cartilage thinner, articular stress is greater in women than men for activities of daily living [12–14].

Hypermobility of joints and its relationship to HOA at the PIP, MCP, and CMC-1 joints also may include a component of sexual dimorphism. Studies have shown a relationship between hypermobility in CMC-1 subluxation and increased radiographic OA in men. For women, there is no evidence of a relationship with hypermobility in PIP, MCP, or CMC-1 joints and radiographic OA [15,16].

Erosive OA is a severe subset of OA and it shares many radiographic features with nonerosive OA [17]. DIP and PIP joints develop erosive OA more often than metacarpal joints [18]. Of patients with classic HOA, 40% have erosive changes, and women are more commonly affected than men. Erosive OA is associated with other diseases, including endocrine disease (hyperparathyroidism, hypothyroidism), microcrystal-related disease (calcium pyrophosphate dihydrate [CPPD] arthroplasty, apatite deposition disease, recurrent calcific periarthritis), chronic renal disease (hemodialysis-associated arthropathy, azotemic osteoarthropathy, amyloid arthropathy), autoimmune disease (scleroderma, Sjögren's syndrome, chronic autoimmune thyroiditis), and frostbite. There also seems to be a hereditary component to erosive HOA [17,19,20].

Risk factors

The onset of OA is mediated by many underlying components. The understanding of OA is changing as it is being linked to a number of risk markers. Genetic, hormonal, overall health, and specific behaviors influence the risk of developing HOA. These factors affect men and women differently. The most important of these aspects are discussed later.

Heredity

Genetics

Genetic components are strong determinants of OA, and the involvement of genes associated with OA is beginning to be clarified. Because OA has multiple possible origins, the genetic contribution is complex. In some instances, genes seem to be operating differently in the two sexes. There may also be different genetic origins for different sites of onset and for the different disease features within a particular site [21].

Support for the genetic contribution to OA onset has come from investigations of sibling pairs. If OA is found in hips and hands in one sibling, there is a threefold incidence of OA in the same locations in the other sibling. Of 191 sibling pairs studied, OA was present in spine (76%), hands (77%), knees (37%), and hips (26%). The most common combinations in probands were spine–hand (59%), spine–knee (27%), and hand–knee (25%) [22,23].

Inheritance patterns and genetic linkages have been confirmed in studies of large population groups and in monozygotic and dizygotic twins. All these studies support genetic linkages and major gene effects in HOA. A relationship between HLA-DRB1*02 (human leukocyte antigens = "transplantation gene") and HOA has been identified and suggests that immunologic responses to low-grade inflammation, as seen in OA, may be restricted [21,24–26].

Sexual dimorphism in the genetics of osteoarthritis

The genetics of OA is one area in which sexual dimorphism is being examined extensively. The genes and their links with other chromosomal sites as they relate to OA are being defined. Studies of a Finnish cohort of 836 monozygotic twins and 1502 dizygotic twins have shown that heritability is greater in women than men. In determining the genetic features that influence the expression of OA, additive gene effects explained 44% of the variance in liability to exhibit the disease, shared familial effects accounted for 20%, and unshared familial effects accounted for 36% in women. Models for genetic effects for men do not fit as well as in women; therefore, the results in men indicate that the genetic effects can be modulated by sex or environmental factors that are distributed differently in men and women. Genome regions and markers used to identify linkage with OA in hip and knee must be examined in HOA [27,28].

OA is a polygenic disease. Polymorphism (ie, DNA variations too common to be explained by mutation) is being studied to detect possible links to certain genes and the onset and progression of OA [29]. Valdes and colleagues [30] identified eight genes associated with OA in women and five such genes associated with men. Single nucleotide polymorphism in two genes, *ADAM12* and *COX2*, were exclusively associated with men and OA. The *ADAM* haplotype CGGT is associated with a reduced risk of OA, perhaps because it results in an arginin residue instead of a glycine residue. The *COX2* gene codes the enzyme cyclo-oxygenase 2, which is the rate-limiting enzyme for prostaglandin E2, known to increase in OA.

Single-nucleotide polymorphism in three genes—*AACT*, *BMP2*, and *ESR1*—are exclusively associated with women and OA. *AACT* encodes α1-antichymotrypsin, a natural inhibitor of a proteinase involved in the degradation of cartilage proteoglycan and is believed to contribute to the reduced risk of knee OA in the Valdes study. *BMP2* encodes the growth factor bone morphogenetic protein 2, which is involved in chondrogenesis and osteogenesis. *ESR1* encodes estrogen receptor α, which, together with estrogen receptor β, is normally present in cartilage of women. Together with interleukin-1β, estrogen is known to modulate proteoglycan degradation and matrix metalloproteinase mRNA expression [30,31].

Genes specific to the hand

Genetic associations between specific anatomic locations of HOA are being established. HOA was examined in the Framingham study because a single entity had not shown a strong linkage between the disease and any chromosomal sites. The data suggested that several chromosomes contain HOA susceptibility genes. The authors concluded that a joint-specific approached may be more rewarding in identifying these linkages than a global approach [32]. For example, human leukocyte antigen class II has been established as a risk factor for OA in the DIP joint [33]. *HFE* gene mutations (HFE is linked to human leukocyte antigen) could be a marker for polyarticular OA phenotype (index and middle finger MCP joints) [34]. Studies by Solovieva and colleagues [35] on a population of Finnish women show that *VDR* gene polymorphism may play a role in the etiology of symmetrical HOA.

In conclusion, knowledge of the specific gene involvement in OA in the hand and elsewhere may lead to improved treatment plans in OA and to the identification of predictive factors [36].

Hormone status

After menopause, women have a more progressive and generalized pattern of OA, causing researchers to study the involvement of estrogen in OA [23,37,38]. Evidence demonstrates the existence of estrogen receptors in articular cartilage. The significance of this as it relates to disease processes is not clear. Young and Stack detected estrogen receptors in the articular cartilage in female and male dogs. Glucocorticoid receptors were also detected in the articular cartilage, but receptors for androgen and progesterone were absent [39]. A study on castrated and adrenalectomized older female baboons injected with 3H-estradiol-17B showed a consistent and heavy uptake of the steroid in biopsies taken 1 hour after the injection [40]. Further studies on sexual dimorphism in articular cartilage have validated that estradiol causes a response in female but not male cultured chondrocytes. The effects noted were an increased DNA synthesis, increased sulfate incorporation, and increased alkaline phosphatase activity [41].

A study evaluating hormone replacement therapy in general (ie, the patients could be using estrogen, combined estrogen/progesterone, alternating estrogen and progesterone, or progesterone alone) showed that such therapy had no influence on the severity or symptoms of HOA [42].

In looking at the onset of OA in the hip, hand, and knee in postmenopausal women, a significantly larger portion of women (4.1%) who used solely estrogen replacement for at least 1 year had hip and hand OA compared with women not using postmenopausal estrogen (PME) (1.1%). When confounding effects of age, body mass index, smoking, exercise, and type of menopause were taken into account, women who used PME were still more likely to have HOA [43]. The Melbourne Women's Midlife health project found that never having used hormone replacement therapy was a risk factor for radiologic hand and knee OA [44].

Hormonal factors may play a role in disease progression and prevention. Estrogen exposure at the time of OA onset may increase the severity of OA in the DIP, whereas breast feeding may be protective for CMC OA [45]. Estrogen has been reported as protective for OA in some studies and linked to the onset of OA in others. The effect of hormones on the disease process of OA is clearly multifactorial. Larger groups and varied populations with multiple factors should be studied.

Osteoporosis

In a bone mineral density study (BMD) of 1779 subjects between the ages of 50 and 96, clinical HOA was present in 6.6% of men and 14.5% of women. The BMD measurements adjusted for age, body mass index, smoking, alcohol, exercise, and current estrogen use were significantly lower only at the hip in patients with HOA versus patients without HOA. Men with HOA had higher multiple-adjusted mean BMD levels at all sites than did men without HOA. The authors of this study concluded that general OA was not associated with increased BMD levels in men or women [46].

Women with radiographic HOA had a significantly increased rate of bone loss at the distal radius than women with normal hand radiographs. Men have no differences in bone loss at any skeletal location when HOA is present. The overall conclusions were that persons with radiographic evidence of OA lose bone at a different rate than patients with normal radiographs do and that this relationship varies with the site of OA and the site at which BMD is measured [47].

There is a direct relationship between DIP OA and CMC OA with osteoporosis as measured by combined cortical thickness and metacarpal index. Consequently, an increase in severity of HOA is associated with an increase severity of osteoporosis. Women with symmetrical DIP OA were found to have an increased risk of developing osteoporosis over time. Symmetrical DIP OA also may be an indicator of generalized OA [48].

Postmenopausal women with erosive HOA had a higher incidence of osteoporosis of the lumbar spine than did postmenopausal women without erosive HOA and controls. This finding suggests a risk for osteoporosis in postmenopausal women with erosive HOA [49]. The presence of HOA does correlate with osteoporosis. This relationship varies for men and women, and the location of the osteoporosis correlates differently in the two sexes. Because of the multifactorial nature of OA, it is not surprising that it has a complex relationship to osteoporosis. This relationship deserves the same scrutiny and research as other aspects of OA.

Presence of concurrent disease

HOA has several associated disease processes. There are significant associations between HOA and hip, knee, and intervertebral disc degeneration. In siblings of probands with OA involvement of hips and hands, there was a threefold incidence of OA in the same locations [23,50]. The Baltimore longitudinal study on aging found an association between HOA and knee OA to be significant. The strength of this association increases with severity of disease, increased number of joints affected, and bilateral disease. Waldron's 1997 study, which examined 115 cases from eighteenth and nineteenth century London, confirmed that HOA and knee OA are associated [51,52]. The presence of HOA at the initial evaluation was a risk factor for future hip and knee OA. The presence of HOA, other OA risk factors, and high OA biomarkers such as type II collagen c-telopeptide degradation product cause an additional risk of future hip and knee OA [10].

In the Framingham study, the strongest predictor for a symptomatic joint in a hand was the presence of OA in the contralateral joint for men and women. This prediction was stronger in women than men. Occurrence of OA in subsequent rows and rays of the hand joints has been documented. Diseases in a joint in a particular row greatly increase the risk for developing OA in the next joint in the same row for men and women. This clustering pattern is also true for rays; DIP OA is strongly associated with the development of PIP OA in the same ray. Chaisson and colleagues [6]noted that female sex was associated with incident OA in the IP joints and the base of the thumb. This association was not apparent between women and the MCP joint.

Age

Increased age may be the major risk factor for OA. For any joint, incidence of OA is 5% for people between the ages of 15 and 44, 25% to 30% for people aged 45 to 64, and between 60% and 90%—depending on population studied—for individuals aged 65 and older [53–57]. Age is strongly correlated with HOA onset in both sexes for all joints [58]. With exceptions, women demonstrated higher HOA prevalence rates for the DIP, third PIP, and first CMC sites [3]. Age and the major gene together are the main sources of interindividual differences in the development of HOA [25].

Weight

Obesity in women has been linked consistently to increased risk for OA, more so than in men. The risk of OA caused by obesity is 9% to 13% per kilogram increase in body weight according to Cicuttini and colleagues [59]. Although many researchers already appreciate obesity as a factor in the onset of knee OA, it is also a risk factor for OA involving the CMC joint. The cause of the relationship between female obesity and OA is unclear. In weight-bearing joints, biomechanical factors may be a direct cause of the onset of OA [60]. Herzog and Fedenco [61] showed that even a single bout of joint loading alters site-specific gene expression of cartilage. Further research on why non–weight-bearing joints of the hand of obese individuals present with OA is warranted [60,62]. Because the pathophysiology of obesity is better understood, studies that aim to explain the linkage between obesity, women, and OA are emerging. The hormone product of the obesity gene, leptin, is manufactured in adipose tissue but also comes from osteoblasts and chondrocytes, which hints at the importance of local production of leptin in joints. A correlation between OA and the presence of leptin in the synovial fluid, cartilage, and osteophytes has been observed; however, few chondrocytes of healthy people produce leptin [63]. Women may have higher levels of leptin than men because of their naturally higher total body fat composition, which may partially explain why OA is more prevalent in women [60]. Fat also stores estrogen, which is another risk factor. A 2003 study of 1476 men and 1519 women born in 1946 showed that increased weight at ages 26, 43, and 53, particularly in combination with low birth weight, was significantly correlated with HOA in men. A correlation between weight and HOA was not found in the women of this cohort [64].

The cortical index has been shown to correlate with adiposity and the index of overall physique, but the BMD was not correlated with physique. Overall physique was determined by physical characteristics, obesity indices, skeletal size, muscular development, and the somatotypes of Heath and Carter in 1967 and Deriabin in 1985 [65].

Nutrition

Low vitamin K intake is associated with increased HOA symptoms. Vitamin K is a cofactor in the production of carboxyglutamic proteins, which are important for cartilage mineralization. The food source of vitamin K, phylloquinone, is found in green leafy vegetables. Low plasma levels of phylloquinone are associated with large osteophytes in hands and knees [66].

Solovieva and colleagues [35] found an association between vitamin D receptor gene polymorphism and symmetrical HOA. These authors suggested that the association between the *VDR* gene and HOA can be altered with calcium intake. Low intake of daily calcium and being a carrier of a haplotype of *VDR* acted together to increase the risk of symmetrical HOA. Low daily calcium intake in individuals who did not carry *VDR* haplotype did not increase HOA incidence.

Summary

Factors other than age and genetics may play a role in explaining the onset of HOA in men and women. Genetic, physiologic, and anatomic differences in men and woman cause the variable expressions of HOA. These different factors affect women's ability to modify HOA before and after the onset of HOA, although it is genetic. By maintaining normal weight, good health, and nutrition, one can diminish the genetic and multifactorial effects of HOA. Future research in genetics, polymorphism, anatomy, hormonal influences, association with other disease processes, and the multifactorial issues will clarify these relationships. Additional studies are needed to investigate the outcomes of gender-specific treatments, joint replacement surgery, and other interventions for HOA.

References

[1] Niu J, Zhang Y, LaValley M, et al. Symmetry and clustering of symptomatic hand osteoarthritis in elderly men and women: the Framingham Study. Rheumatology 2003;42:343–8.

[2] Cushnaghan J, Dieppe P. Study of 500 patients with limb joint osteoarthritis. I. Analysis by age, sex, and distribution of symptomatic joint sites. Ann Rheum Dis 1991;50:8–13.

[3] Wilder F, Barrett J, Farina E. Joint-specific prevalence of osteoarthritis of the hand. Osteoarthritis Cartilage 2006;14(9):953–7.

[4] Lawrence J, Bremner J, Biers F. Osteoarthritis: prevalence in the population and relationship between symptoms and x-ray changes. Ann Rheum Dis 1966;25:1–5.

[5] Srikanth V, Fryer J, Zhai G, et al. A meta-analysis of sex differences prevalence, incidence and severity of osteoarthritis. Osteoarthritis Cartilage 2005;13(9): 769–81.

[6] Chaisson C, Zhang Y, McAlidon T, et al. Radiographic hand osteoarthritis: Incidence, patterns, and influence of pre-existing disease in a population based sample. J Rheumatol 1997;24(7):1337–43.

[7] Poole J, Sayer A, Hardy R, et al. Patterns on interphalangeal hand joint involvement of osteoarthritis among men and women: a British cohort study. Arthritis Rheum 2003;48(12):3371–6.

[8] Kessler S, Stove J, Puhl W, et al. First carpometacarpal and interphalangeal osteoarthritis of the hand in patients with advanced hip or knee OA: are there differences in the aetiology? Clin Rheumatol 2003; 22(6):409–13.

[9] Haara M, Manninen P, Kroger H, et al. Osteoarthritis of finger joints in Finns aged 30 or over: prevalence, determinants, and association with mortality. Ann Rheum Dis 2003;62:151–8.

[10] Dahaghin S, Bierma-Zeinstra S, Ginai A, et al. Prevalence and determinants of one month hand pain and hand related disability in the elderly (Rotterdam study). Ann Rheum Dis 2005;64:99–104.

[11] Rossignol M, Leclerc A, Hilliquin P, et al. Primary osteoarthritis and occupations: a national cross sectional survey of 10412 symptomatic patients. Occup Environ Med 2003;60:882–6.

[12] Cope J, Berryman A. Martin, et al. Robusticity and osteoarthritis at the trapeziometacarpal joint in a Bronze Age population from Tell Abraq, United Arab Emirates. Am J Phys Anthropol 2005;126(4): 391–400.

[13] Xu L, Strauch R, Ateshian G, et al. Topography of the osteoarthritic thumb carpometacarpal joint and its variations with regard to gender, age, site and osteoarthritic stage. J Hand Surg [Am] 1998;23(3): 454–64.

[14] Kovler M, Lundon K, McKee N, et al. The human first carpometacarpal joint: osteoarthritic degeneration and 3-dimensional modeling. J Hand Ther 2004; 17(4):393–400.

[15] Kraus V, Li Y, Martin E, et al. Articular hypermobility is a protective factor for hand osteoarthritis. Arthritis Rheum 2004;50(7):2178–83.

[16] Hunter D, Zhang Y, Sokolove J, et al. Trapeziometacarpal subluxation predisposes to incident trapeziometacarpal osteoarthritis (OA): the Framingham Study. Osteoarthritis Cartilage 2005;13:953–7.

[17] Punzi L, Ramonda R, Sfriso P, et al. Erosive osteoarthritis. Best Pract Res Clin Rheumatol 2004;18(5): 739–58.

[18] Verbruggen G, Veys E. Erosive and non-erosive hand osteoarthritis: use and limitations of two scoring systems. Osteoarthritis Cartilage 2000;8(Suppl A): S45–54.

[19] Cavasin F, Punzi L, Ramonda R, et al. Prevalence of erosive osteoarthritis of the hand in a population from Venetian area. Reumatismo 2004;56(1):46–50.

[20] Ehrlich GE. Erosive osteoarthritis: presentation, clinical pearls, and therapy. Curr Rheumatol Rep 2001;3(6):484–8.

[21] Spector T, MacGregor A. Risk factors for osteoarthritis: genetics. Osteoarthritis Cartilage 2004; 12(Suppl A):S39–44.

[22] Riyazi N, Meulenbelt I, Kroon H, et al. Evidence for familial aggregation of hand, hip, and spine but not knee osteoarthritis in siblings with multiple joint involvements: the GARP study. Ann Rheum Dis 2005; 64:438–43.

[23] Peat G, Muelenbelt I, Kroon H, et al. Clinical assessment of the osteoarthritis patient. Best Pract Res Clin Rheumatol 2001;15(4):527–44.

[24] Jonsson H, Manolescu I, Stefansson E, et al. The inheritance of hand osteoarthritis in Iceland. Arthrits Rheum 2003;48(2):391–5.

[25] Livshits G, Kalichman L, Cohen Z, et al. Mode of inheritance of hand osteoarthritis in ethnically homogenous pedigrees. Hum Biol 2002;74(6):849–60.

[26] Moos V, Menard J, Sieper J, et al. Association of HLA-DRB1*02 with osteoarthritis in a cohort of 106 patients. Rheumatology (Oxford) 2002;41(6): 666–9.

[27] Kaprio J, Kujala U, Peltonen L, et al. Genetic liability to osteoarthritis may be greater in women and men. BMJ 1996;313:232.

[28] Loughlin J. Stratification analysis of an osteoarthritis genome screen: suggestive linkage to chromosome 4, 6, and 16 [letter to the editor]. Am J Hum Genet 1999;65:1795–8.

[29] Chernajovski Y, Winyard P, Kabourisi P. Advances in understanding the genetic basis of rheumatoid arthritis and osteoarthritis: implications for therapy. Am J Pharmacogenomics 2002;2(4):223–34.

[30] Valdes A, Van Oene M, Hart D, et al. Reproducible genetic associations between candidate genes and clinical knee osteoarthritis in men and women. Arthritis Rheum 2006;54(2):533–9.

[31] Richette P, Corvol M, Bardin T. Estrogens, cartilage and osteoarthritis. Joint Bone Spine 2003;70:257–62.

[32] Hunter D, Demissie S, Cupples L, et al. A genome scan for joint-specific hand osteoarthritis susceptibility: the Framingham Study. Arthritis Rheum 2004;50(8):2489–96.

[33] Riyazi N, Spee J, Huizinga TW, et al. HLA class II is associated with distal interphalangeal osteoarthritis. Ann Rheum Dis 2003;62(3):227–30.

[34] Carroll G. HFE gene mutations are associated with osteoarthritis in the index or middle finger metacarophalangeal joints. J Rheumatol 2006;33(4):741–3.

[35] Solovieva S, Hirvonen A, Siivola P, et al. Vitamin D receptor gene polymorphisms and susceptibility of hand osteoarthritis in Finnish women. Arthritis Res Ther 2006;8:R20.

[36] Ivkovic A, Pascher A, Hudgetz D, et al. Current concepts in gene therapy of the musculoskeletal system. Acta Chir Orthop Traumatol Cech 2006;73(2): 115–22.

[37] Moskowitz RW. Clinical and laboratory findings in osteoarthritis. In: Hollander J, McCarty D Jr, editors. Arthritis and allied conciliations. Philadelphia: Lea and Febiger; 1972.

[38] Kellgren J, Lawrence J, Bier F. Genetic factors in generalized osteoarthrosis. Ann Rheum Dis 1963; 22:237–45.

[39] Young P, Stack M. Estrogen and glucocorticoid receptors in adult canine articular cartilage. Arthritis Rheum 1982;25(5):568–73.

[40] Sheridan P, Aufdemorte T, Holt G, et al. Cartilage of the baboon contains estrogen receptors. Rheumatology 1985;5(6):279–81.

[41] Kinney R, Schwartz Z, Week K, et al. Human articular chondrocytes exhibit sexual dimorphism in their response to 17β-estradiol. Osteoarthritis Cartilage 2005;13:330–7.

[42] Maheu E, Dreiser R, Guillou G, et al. Hand osteoarthritis patients characteristics according to the existence of a hormone replacement therapy. Osteoarthritis Cartilage 2000;8(Suppl A):S33–7.

[43] Von Muhlen D, Morton D, Von Muhlen C. Postmenopausal estrogen and increased risk of clinical osteoarthritis at the hip, hand, and knee in older women. J Womens Health Gend Based Med 2002; 11(6):511–8.

[44] Szoeke C, Cicuttini F, Guthrie J, et al. Factors affecting the prevalence of osteoarthritis in healthy middle-aged women: data from the longitudinal Melbourne Women's Midlife Health Project. Bone 2006;39(5):1149–55.

[45] Cooley H, Stankovich J, Jones G. The association between hormonal and reproductive factors and

hand osteoarthritis. Maturitas, the European Menopause Journal 2003;45:257–65.
[46] Schneider D, Barrett-Connor E, Morton D. Bone mineral density and clinical osteoarthritis in elderly men and women: the Rancho Bernardo study. J Rheumatol 2002;29(7):1467–72.
[47] Hichberg M, Lethbridge-Ceujku M, Tobin J. Bone mineral density and osteoarthritis: data from the Baltimore Longitudinal study of aging. Osteoarthritis Cartilage 2004;12(Suppl A):S45–8.
[48] Haara M, Arokoski J, Kroger H, et al. Association of radiological hand osteoarthritis with bone mineral mass: a population study. Rheumatology (Oxford) 2005;44(12):1549–54.
[49] Zoli A, Lizzio M, Capuano A, et al. Osteoporosis and bone metabolism in post menopausal women with osteoarthritis of the hand. Menopause 2006; 13(3):462–6.
[50] Riyazi N, Muelenbelt I, Kroon H, et al. Evidence for familial aggregation of hand, hip, and spine but not knee osteoarthritis in siblings with multiple joint involvements: the GARP study. Ann Rheum Dis 2005; 64:438–43.
[51] Hirsch R, Lethbridhe-Ceijku M, Scott W, et al. Association of hand and knee osteoarthritis: evidence for a polyarticular disease subset. Ann Rheum Dis 1996;55(1):25–9.
[52] Waldron H. Association between osteoarthritis of the hand and knee in a population of skeletons from London. Ann Rheum Dis 1997;52(2):116–8.
[53] Dieppe P. the classification and diagnosis of osteoarthritis. In: Kuettner KE, Goldberg V, editors. Osteoarthritic disorders. Rosemont (IL): American Academy of Orthopaedic Surgeons; 1995. p. 5–12.
[54] Elders M. The increasing impact of arthritis on public health. J Rheumatol Suppl 2000;60:6–8.
[55] Felson D. The epidemiology of osteoarthritis: prevalence and risk factors. In: Kuettner KE, Goldberg V, editors. Osteoarthritic disorders. Rosemont (IL): American Academy of Orthopaedic Surgeons; 1995. p. 13–24.
[56] Praemer A, Furner S, Rice D. Musculoskeletal conditions in the United States. Rosemont (IL): American Academy of Orthopaedic Surgeons; 1999. p. 182.
[57] Yelin E. The economics of osteoarthritis. In: Brandt K, Doherty M, Lohmander L, editors. Osteoarthritis. Oxford: Oxford University Press; 1998. p. 23–30.
[58] Kalichman L, Cohen Z, Kobyliansky E, et al. Pattern of joint distribution in hand osteoarthritis: contribution of age, sex and handedness. Am J Hum Biol 2004;16:125–34.
[59] Cicuttini F, Baker J, Spector T. The association of obesity with osteoarthritis of the hand and knee in women: a twin study. J Rheumatol 1996;23(7): 1221–6.
[60] Teichtahl A, Wluka AE, Proietto J, et al. Obesity and the female sex, risk factors for knee osteoarthritis that may be attributable to systemic or local leptin biosynthesis and its cellular effects. Med Hypotheses 2005;65(2):312–5.
[61] Herzog W, Federico S. Considerations on joint and articular cartilage mechanics. Biomech Model Mechanobiol 2006;5(2–3):64–81.
[62] Hart D, Spector T. The relationship of obesity, fat distribution and osteoarthritis in women in the general population: the Chingford Study. J Rheumatol 1993;23(7):1221–6.
[63] Dumond H, Presle N, Terlain B, et al. Evidence for a key role of leptin in osteoarthritis. Arthritis Rheum 2003;48(11):3118–29.
[64] Sayer A, Poole J, Cox V, et al. Weight from birth to 53 years: a longitudinal study of the influence on clinical hand osteoarthritis. Arthritis Rheum 2003; 48(4):1030–3.
[65] Kalichman L, Malkin I, Kobylinsky E. Association between physique characteristics and hand skeletal aging status. Am J Phys Anthropol 2005;128(4): 889–95.
[66] Neogi T, Booth S, Zhang Y, et al. Low vitamin K status is associated with osteoarthritis in the hand and knee. Arthritis Rheum 2006;54(4):1255–61.

Sexual Dimorphism in Degenerative Disorders of the Spine

Neil A. Manson, MD, FRCSC, Edward J. Goldberg, MD, Gunnar B.J. Andersson, MD, PhD*

Department of Orthopaedic Surgery, Rush University Medical Center, 1725 West Harrison Street, Suite 1063, Chicago, IL 60612, USA

The female predominance observed in adolescent idiopathic scoliosis is well demonstrated in the orthopaedic literature. The female/male ratio for curves more than 10° is 1.4:1, yet for curves more than 30° it is 10:1 [1,2]. Progression is more likely in women. Discrepancies attributed to sex in degenerative disorders of the spine are less frequently reported. These discrepancies depend on spinal region and disorder and may be multifactorial, including developmental, anatomic, biomechanical, and environmental. This article reviews some differences in degenerative disorders of the spine potentially related to sexual dimorphism.

Spinal development: embryology, maturation, and anatomy

Vertebral column development begins at 4 weeks' gestation as mesenchymal cells move into three main areas: around the notochord, surrounding the neural tube, and in the body wall. Around the notochord some densely packed cells form the intervertebral disc, whereas others fuse to form the mesenchymal centrum of the vertebra. The notochord degenerates and disappears where it is surrounded by the vertebral body, but between the vertebrae, the notochord expands to form the nucleus pulposus. The mesenchymal cells surrounding the neural tube form the vertebral arch, and cells in the body wall eventually develop into ribs [3].

During week 6 of gestation, chondrification centers appear in each mesenchymal vertebra. A cartilaginous centrum, vertebral arch, and spinous and transverse processes develop. At week 8, three primary centers of ossification are evident: one in the centrum and one in each half of the vertebral arch. The literature suggests sex differences during these early stages of spinal development. Prenatally the ossification timing was noted to be significantly earlier in girls than boys [4]. At birth each vertebra consists of three bony sections connected by cartilage. The halves of the vertebral arch fuse and the arch fuses with the centrum at 3 to 6 years of age [3].

Vertebral bodies grow horizontally by the process of periosteal ossification, with the largest changes occurring in the first 7 years after birth. During this period, the lateral diameter increases up to five times and the anteroposterior diameter increases up to ten times. Horizontal growth continues to increase after puberty by 5% to 10%. Vertical growth is rapid during childhood as vertebral body height increases two to four times in the first 5 years after birth. Fusion of the ring apophyses occurs between age 14 and 21 and signals the end of longitudinal growth [5].

After puberty, five secondary centers of ossification appear: tips of the spinous and transverse processes and two rim epiphyses. Secondary centers unite with the rest of the vertebra around age 25 years.

Several authors describe sexual dimorphism in relation to spinal anatomy [6–11]. Taylor and Twomey [11] extensively analyzed the vertebral column during the growing years. At each annual stage up to age 9, spine lengths were almost

* Corresponding author.
E-mail address: gunnar_andersson@rsh.net (G.B.J. Andersson).

identical in boys and girls. From 9 to 12 years, height changes were 61% greater in girls than boys. The growth spurt, which began at age 8.5 years in girls, occurred at 13.5 years in boys. During this time the height/transverse diameter index differences resulted from a greater growth in vertical height in girls and greater growth in transverse diameter in boys. Although female vertical growth progressed more rapidly, increasing male hormones produced increases in muscle bulk and strength. Horizontal growth depends partly on mechanical influences of these stronger muscles, and greater horizontal growth was observed during this time in boys. Once the male growth spurt began, vertebrae of larger transverse diameter began to increase in height. The final result was a slightly shorter but significantly more slender female vertebral column [11].

Schultz and colleagues [12] corroborated these results but found the female vertebral column to be more slender at all ages from 9 to 16 years. At the younger ages, the slenderness was caused by greater vertebral column height in girls, with similar frontal and sagittal diameters noted between genders. In the older ages, the slenderness was caused by smaller frontal and sagittal diameters in girls compared with boys. The spine morphology was linked to the earlier vertical growth spurt and achievement of maturity in girls [12]. Researchers speculated that the more slender female vertebral column would be more prone to deforming forces, which contribute to the increased risk of progression of scoliosis.

The cervical spine also shows significant gender differences. Although the vertebral body height and anteroposterior measurements are greater in boys, no significant differences have been noted in canal diameter. The anteroposterior vertebral body size/canal size ratio—the so-called "Torg ratio" [13]—was found to be greater in girls at all cervical levels [6–10].

With aging, degenerative spine changes appear in individuals of both genders. The biomechanics of spinal motion progress through early dysfunction, followed by instability, and finally regained stability, because of the biomechanical changes caused by disc degeneration, facet joint arthritis, muscle alterations, and ligament degeneration [14]. This progression has been confirmed and has demonstrated sex specificity in biomechanical cadaver testing. Female specimens demonstrated greater overall range of motion in all planes regardless of severity of degeneration. The progression of degenerative change correlated with motion changes, and these changes were similar between genders. Although significance was observed in all motion planes in male specimens, only axial rotation was significantly affected in female segments [15].

Finally, spinal cord development proceeds concomitantly with vertebral growth. At 6 months' gestation, the end of the spinal cord lies at S1. In the newborn it lies at L3, and in the adult it lies at L2 [3]. A recent MRI investigation suggested that although there is no difference in thecal sac termination, the conus medullaris ends slightly lower in women than men. On average, in women the conus ended at the middle-lower region of the middle third of L1 and in men it ended at the lowest part of the upper third of L1 [16].

Qualitative development: density

Bone mass increases during the growth period and peaks by young adulthood. Although the greatest gain in bone mass takes place during the growth spurt of adolescence, bone mineral density continues to increase for several years afterward [17].

Sex is a well-known factor for determining bone mass and osteoporosis risk. Although bone mass in young adult men is greater than in women, the sex-related differences seem to depend more on bone size rather than density. When height adjusted, the volumetric peak bone mineral density is similar in both sexes during early adulthood. Greater values for bone size and mineral content in men is related only to their greater overall size in general [17,18].

The concerns regarding loss of bone mineral density and subsequent osteoporosis observed with aging are often associated with women. Elderly men are felt to possess greater bone size and bone mineral content, which has not proven to be the case in the vertebral column. In fact, correction for height and weight may confer a lower bone mineral density in men rather than women [19,20]. The deterioration of bone mineral content with age has been observed to be the result of poor bone deposition in elderly women compared with men, because both sexes demonstrated similar rates of bone resorption [21].

The sex differences in the vertebral column with aging pertain to bone size, density, and remodeling. The interplay of these factors may alter the progression of degenerative spinal disorders in one sex relative to another. Larger absolute vertebral size and density may afford protection from degeneration. The significantly greater amount of bone deposition versus absorption seen in elderly men may confer rebuilding properties to the aging spine, for example.

Pregnancy

Some degree of musculoskeletal discomfort is inherent during pregnancy. The incidence of low back pain in pregnancy approximates 50%. Physiologic changes include fluid retention and edema, ligamentous laxity, weight gain, and hyperlordosis. These changes contribute to increase the mechanical strain on the lower back.

An extensive prospective analysis of 855 pregnant Swedish women provided many insights into pregnancy and back pain. A past history of back pain correlated with occurrence, intensity, and duration of back pain during pregnancy [22]. Back pain was documented in 49% of these women during pregnancy, correlating to maternal age, multiparity, back pain in the past, a poor self-concept of back health, and physical work [23]. Back pain persisted, because 37% of the women reported posterior pelvis or lumbar pain at 12 months postpartum. Sick leave during pregnancy, physical work, previous back pain, and menstrual pain correlated with persistent symptoms [24]. Biomechanical factors related to back pain in pregnancy included sagittal and transverse abdominal diameters and lumbar lordosis [25]. It was felt that back pain cannot be explained simply by biomechanical factors, however [23]. No etiology was offered by these epidemiologic studies, and the relationship of pain to subsequent degeneration has not been studied.

Research has suggested that pregnancy may be a factor for the development of degenerative spondylolisthesis in susceptible women. Women with previously diagnosed spondylolisthesis, however, demonstrate no increase in slippage during pregnancy [26]. The effect of physiologic changes observed during pregnancy on the development of later degenerative disease has not been evaluated.

Degenerative spine disorders demonstrating sexual dimorphism

Degenerative disorders of the spine are symptomatic manifestations of the normal aging process. In men, back pain increases up to age 50 years and then declines; in women, the peak is observed at 60 years [27]. Sexual dimorphism is apparent in degenerative disorders of the spine, although no disease entity is exclusive to either sex. Degenerative disorders of the lumbar spine, for example, are usually first identified in patients in their fourth decade of life and occur with equal frequency in men and women [28]. Of these disorders, however, women demonstrate a greater incidence of instability-based diseases. In contrast, men show greater incidence of disease caused by structural degeneration.

Male predominance

Structural deterioration leads to symptomatic degenerative disorders more often in men. These observations have been made in all regions of the spine. Isthmic spondylolisthesis is observed approximately twice as often in men as in women. The prevalence is not age related, and no cause has been elucidated [29].

Symptomatic spinal stenosis is more prevalent in cervical and lumbar regions in men. Several authors have noted differences in relative canal diameter in the cervical spine in men compared with women [7–9]. This difference may be the reason for the higher rate of cervical myelopathy observed in men [6]. Similar findings are reported in the lumbar spine, because lumbar stenosis (without degenerative spondylolisthesis) is observed twice as often in men as women. Again, the relatively smaller size of the normal spinal canal at the third to fifth lumbar levels in men is believed to be a causative factor [30,31].

Disc degeneration occurs earlier and with greater severity in men. More severe disc degeneration was observed during autopsy at nearly all ages in the male lumbar spine [32]. In children and adults, symptomatic lumbar disc herniation is observed twice as frequently in male subjects [7,33,34]. Thoracic disc herniations are also slightly more common in men [35,36]. In contrast, no sex difference has been reported in herniations of the cervical disc [7].

Ankylosing spondylitis affects men more often than women, with a ratio range of 3:1 to 10:1. Although both genders display a similar clinical picture, the age of onset is earlier and the disease severity is greater in men than in women. No reason has been identified for this difference, but hormonal effects have been suggested [37,38].

Female predominance

Issues of stability result in symptomatic degeneration in the female spine more often than in the male spine. Ligament laxity secondary to hormonal influences is implicated in the low back pain of pregnancy [26]. The long-term influence of these transient physiologic changes has received little attention.

Degenerative spondylolisthesis develops predominantly between the fourth and fifth lumbar vertebrae. Prevalence is correlated with age, with 10% of women over age 60 demonstrating radiographic changes. Cadaveric investigation revealed listhesis rates five times more frequently in women than men. One study correlated spondylolisthesis with a history of pregnancy, speculating that increased soft tissue relaxation and joint laxity, a large flexion moment on the lumbar spine, and abdominal muscle weakness are causative factors [30,39].

Rheumatoid arthritis affects women 2.5 times more often than men [40]. Atlantoaxial instability is the most common cervical pathology noted in patients who have rheumatoid arthritis, occurring in 50% of patients with symptomatic cervical changes. Subluxation results from attenuation of the transverse atlantoaxial ligament and the apical ligaments secondary to the inflammatory process. Basilar invagination often follows because of destruction of the lateral masses and occipital condyles. This type of subluxation is present in 40% of symptomatic patients and is believed by some to be a continuum of the atlantoaxial destabilizing process. Subaxial subluxation is present in 20% of patients and occurs because of destruction of facets, ligaments, and the intervertebral discs. Although women suffer the disease at greater rates than men, some researchers have suggested that male sex may be correlated with more extensive cervical involvement [41].

Perioperative considerations

Surgical intervention for spinal pathology is influenced by sex [34]. Katz and colleagues [42] suggested that women undergoing laminectomy for lumbar spinal stenosis had worse preoperative functional status and perhaps more advanced disease. Several reasons were offered, including the following: adversity to risk taking by women, practical considerations, including caring for a disabled spouse, subtle barriers to referral and access, and perception of greater disability by women when performing specific tasks. Fortunately, women demonstrated similar surgical success and greater improvement in functional status postoperatively compared with men [42].

A similar analysis of elderly patients undergoing decompression surgery for lumbar spinal stenosis offered corroborating evidence. Postoperative satisfaction correlated directly to preoperative expectations. Preoperative expectations were more positive in patients who were older, male, not living alone, and had more years of education. Postsurgical satisfaction was lower in women than men despite equivalent outcomes. The cause of the discrepancy was felt to be multifactorial and not related to the surgical intervention itself [43,44].

Summary

Sexual dimorphism is evident during formation, growth, and development of the spine. Pregnancy alters spine physiology. The subsequent processes of aging and spinal degeneration adversely affect men and women slightly differently. Although degenerative changes are observed at similar rates in both sexes, women seem to be more susceptible to degenerative changes that lead to instability and malalignment, such as spondylolisthesis. Men, however, suffer to a greater extent from structural deterioration, such as stenosis or disc degeneration. Surgical satisfaction is greater in men and may be attributed to poorer preoperative function and lower expectations in women.

References

[1] Lenke L, Dobbs M. Idiopathic scoliosis. In: Frymoyer J, Wiesel S, editors. The adult and pediatric spine. 3rd edition. Philadelphia: Lippincott Williams Wilkins; 2004. p. 337–60.
[2] Wenger D, Mubarak S, Chambers H, et al. Pediatric scoliosis. In: Garfin S, Vaccaro A, editors. Orthopaedic knowledge update: spine. Rosemont (IL): AAOS; 1997. p. 183–94.
[3] Moore K, Persuad T. The developing human: clinically oriented embryology. 7th edition. Toronto: WB Saunders; 2003.
[4] Vignolo M, Ginocchio G, Parodi A, et al. Fetal spine ossification: the gender and individual differences illustrated by ultrasonography. Ultrasound Med Biol 2005;31(6):733–8.
[5] Vaccaro A. Spine anatomy. In: Garfin S, Vaccaro A, editors. Orthopaedic knowledge update: spine. Rosemont (IL): AAOS; 1997. p. 3–17.
[6] Hukuda S, Kojima Y. Sex discrepancy in the canal/body ratio of the cervical spine implicating the prevalence of cervical myelopathy in men. Spine 2002;27(3):250–3.
[7] Kelley L. In neck to neck competition are women more fragile? Clin Orthop 2000;372:123–30.
[8] Liguoro D, Vandermeersch B, Guerin J. Dimensions of cervical vertebral bodies according to age and sex. Surg Radiol Anat 1994;16(2):149–55.
[9] Lim J, Wong H. Variation of the cervical spinal Torg ratio with gender and ethnicity. Spine J 2004;4(4):396–401.

[10] Tatarek N. Variation in the human cervical neural canal. Spine J 2005;5(6):623–31.
[11] Taylor J, Twomey L. Sexual dimorphism in human vertebral body shape. J Anat 1984;138(2):281–6.
[12] Schultz A, Sorensen S, Andersson G. Measurements of spine morphology in children, ages 10–16. Spine 1984;9(1):70–3.
[13] Torg J, Pavlov H, Genuario S, et al. Neuropraxia of the cervical spinal cord with transient quadriplegia. J Bone Joint Surg Am 1986;68A:1354–70.
[14] Kirkaldy-Willis W, Farfan H. Instability of the lumbar spine. Clin Orthop 1982;165:110–23.
[15] Fujiwara A, Lim T, An H, et al. The effect of disc degeneration and facet joint osteoarthritis on the segmental flexibility of the lumbar spine. Spine 2000;25(23):3036–44.
[16] Soleiman J, Demaerel P, Rocher S, et al. Magnetic resonance imaging study of the level of termination of the conus medullaris and the thecal sac: influence of age and gender. Spine 2005;30(16):1875–80.
[17] Valero C, Zarrabeitia M, Hernandez J, et al. Bone mass in young adults: relationship with gender, weight and genetic factors. J Intern Med 2005;258:554–62.
[18] Wu X, Yang Y, Zhang H, et al. Gender differences in bone density at different skeletal sites of acquisition with age in Chinese children and adolescents. J Bone Miner Metab 2005;23(3):253–60.
[19] Nieves J, Formica C, Ruffing J, et al. Males have larger skeletal size and bone mass than females, despite comparable body size. J Bone Miner Res 2005;20(3):529–35.
[20] Tuck S, Pearce M, Rawlings D, et al. Differences in bone mineral density and geometry in men and women: the Newcastle thousand families study at 50 years old. Br J Radiol 2005;78(930):493–8.
[21] Duan Y, Turner C, Kim B, et al. Sexual dimorphism in vertebral fragility is more the result of gender differences in age-related bone gain than bone loss. J Bone Miner Res 2001;16(12):2267–75.
[22] Ostgaard H, Andersson G. Previous back pain and risk of developing back pain in a future pregnancy. Spine 1991;16(4):432–6.
[23] Ostgaard H, Andersson G, Karlsson K. Prevalence of back pain in pregnancy. Spine 1991;16(5):549–52.
[24] Ostgaard H, Andersson G. Postpartum low-back pain. Spine 1992;17(1):53–5.
[25] Ostgaard H, Andersson G, Schultz A, et al. Influence of some biomechanical factors on low-back pain in pregnancy. Spine 1993;18(1):61–5.
[26] Borg-Stein J, Dugan S, Gruber J. Musculoskeletal aspects of pregnancy. Am J Phys Med Rehabil 2005;84:180–92.
[27] Burdorf A. Exposure assessment of risk factors for disorders of the back in occupational epidemiology. Scand J Work Environ Health 1992;18:1–9.
[28] Burkus J, Zdeblick T. Lumbar disc disease: pathophysiology of lumbar spondylosis and discogenic back pain. In: Wiesel JFS, editor. The adult and pediatric spine. 3rd edition. Philadelphia: Lippincott Williams & Wilkins; 2004. p. 899–911.
[29] Virta L, Ronnemaa T, Osterman K, et al. Prevalence of isthmic lumbar spondylolisthesis in middle-aged subjects from eastern and western Finland. J Clin Epidemiol 1992;45(8):917–22.
[30] Lauerman W, Goldsmith M. Spine. In: Miller M, editor. Review of orthopaedics. 3rd edition. Philadelphia: WB Saunders; 2000. p. 353–78.
[31] Truumees E. Spinal stenosis: pathophysiology, clinical and radiologic classification. Instr Course Lect 2005;54:287–302.
[32] Miller J, Schmatz C, Schultz A. Lumbar disc degeneration: correlation with age, sex, and spine level in 600 autopsy specimens. Spine 1988;13(2):173–8.
[33] McLain R. Lumbar disc disease: lumbar disc disease in children. In: Wiesel JFS, editor. The adult and pediatric spine. 3rd edition. Philadelphia: Lippincott Williams & Wilkins; 2004. p. 929–33.
[34] Andersson G. The epidemiology of spinal disorders. In: Frymoyer J, editor. The adult spine: principles and practice. 2nd edition. Philadelphia: Lippincott-Raven; 1997. p. 93–141.
[35] Belanger T, Emery S. Thoracic disc disease and myelopathy. In: Wiesel JFS, editor. The adult and pediatric spine. Philadelphia: Lippincott Williams & Wilkins; 2004. p. 855–64.
[36] Vanichkachorn J, Vaccaro A. Thoracic disk disease: diagnosis and treatment. J Am Acad Orthop Surg 2000;8:159–69.
[37] Jimenez-Balderas F, Mintz G. Ankylosing spondylitis: clinical course in women and men. J Rheumatol 1993;20(12):2069–72.
[38] Khan M. Ankylosing spondylitis. In: Klippel J, editor. Primer on the rheumatic diseases. 11th edition. Atlanta (GA): Arthritis Foundation; 1997. p. 189–93.
[39] Lin K, Jenis L. Degenerative lumbar spondylolisthesis. Seminars in Spine Surgery 2003;15(2):150–9.
[40] Goronzy J, Weyand C. Rheumatoid arthritis: epidemiology, pathology, and pathogenesis. In: Klippel J, editor. Primer on the rheumatic diseases. 11th edition. Atlanta (GA): Arthritis Foundation; 1997. p. 155–61.
[41] Oxner W, Kang J. Inflammatory arthritis of the spine. In: Koval K, editor. Orthopaedic knowledge update 7. Rosemont (IL): AAOS; 2002. p. 689–701.
[42] Katz J, Wright E, Guadagnoli E, et al. Differences between men and women undergoing major orthopaedic surgery for degenerative arthritis. Arthritis Rheum 1994;37(5):687–94.
[43] Gepstein R, Arinzon Z, Adunsky A, et al. Decompression surgery for lumbar spinal stenosis in the elderly: preoperative expectations and postoperative satisfaction. Spinal Cord 2005;43:1–5.
[44] Shabat S, Folman Y, Arinzon Z, et al. Gender differences as an influence on patients' satisfaction rates in spinal surgery of elderly patients. Eur Spine J 2005;14(10):1027–32.

Sexual Dimorphism in Adolescent Idiopathic Scoliosis

Cathleen L. Raggio, MD

Hospital for Special Surgery, 535 East 70th Street, New York, NY 10021, USA

Adolescent idiopathic scoliosis (AIS), defined as a structural lateral curvature of the spine measuring 10° or greater on a posteroanterior spine radiograph and associated with a clinical rotational deformity [1] in an individual 11 years of age or older, is one of the orthopedic disorders in which clinical evidence of sexual dimorphism is most marked. Although the female-to-male ratio for curves greater than 10° is nearly equal at 1.4:1, the ratio for curves greater than 30° approaches 10:1 [2,3].

The cause of AIS is unknown but generally believed to be multifactorial. In addition, because of the female preponderance of patients requiring surgery, many studies exclude boys from their analysis or contain so few as to be statistically irrelevant. The available data, however, suggest that there are important differences in male and female AIS that impact diagnosis, treatment, and outcomes.

Epidemiology/genetic influences

Population studies indicate that 11% of first-degree relatives are affected compared with only 2.4% of second-degree and 1.4% of third-degree relatives [4]. Similarly, a meta-analysis of 68 sets of twins who had scoliotic curves showed a prevalence of scoliosis in 73% of the monozygous twins compared with 36% of the dizygotic twins [5]. The mode of inheritance is currently a source of active debate [4,6–12].

Cowell and colleagues [13] postulated the presence of an X-linked trait based on their analysis of a single large family in which there was no evidence of male-to-male transmission. Miller and colleagues' [10] later work found evidence for and against this mode of transmission. One of the most powerful arguments against this being an X-linked trait is that boys are typically more commonly affected than girls in X-linked disorders, not less [14].

Growth and development

A concise summary of spine growth and development is provided by Manson and colleagues elsewhere in this issue. In brief, as demonstrated by Taylor and Twomey [15], spine growth is virtually identical in boys and girls until age 9 years. From age 9 to 12 years, height changes were 61% greater in girls than in boys. The growth spurt, which begins at age 8.5 years in girls, occurs much later (age 13.5 years) in boys. During the growth spurt years, a significant difference in the height/transverse diameter index results from a greater growth in vertical height in girls and a greater growth in transverse diameter in boys. These growth differences ultimately result in a slightly shorter but significantly more slender female vertebral column. Some investigators have speculated that the more slender female vertebral column is more prone to deforming forces, thus possibly contributing to the increased risk of progression of scoliosis and the higher number of girls requiring surgery.

Diagnosis

Clinical presentation

AIS is generally first noted when the pediatrician, school screener, or parent notes an asymmetry of the trunk or shoulders or "uneven hips." Girls typically present around age 11 to 14 years, whereas boys present later, at age 12 to 15 years. Most researchers believe that this occurs because

E-mail address: raggioc@hss.edu

growth velocity patterns are sex specific [1]. Girls generally begin their adolescent growth spurt at the onset of puberty, whereas boys may be in an advanced stage of puberty when their adolescent growth spurt begins [16]. Sucato and colleagues found that boys and girls present with primary coronal curves of similar magnitude [18].

Curve pattern

Curves are defined by the anatomic area of the spine in which the apex of the curve occurs and by the direction of the curve (ie, right or left). The three most common curves are thoracic, thoracolumbar (apex T12-L1), and lumbar. Curves may be single (a single area involved) or double (two areas involved). The most common curve is the right thoracic with compensatory left lumbar curve. Left thoracic curves are frequently associated with an underlying neurologic condition. Some studies have reported sex-related differences in left curves. More boys had left curves, and girls who were premenarchal at diagnosis had a higher incidence of left curves than older girls [19,20].

Curve stiffness

Spine stiffness may also contribute susceptibility to increased curve progression and, theoretically, make bracing less effective. The literature on this topic is contradictory. Most investigators report that girls have stiffer spines than boys on initial presentation [21]; however, because brace treatment fails to halt curve progression in boys and because boys typically have larger curves at surgery than girls, other investigators have reported stiffer curves in boys [18,22].

Body habitus

Body habitus has been postulated as a risk factor for scoliosis [23]. Many investigators have noted that spine slenderness and ectomorphy allow the development of a curvature in response to biomechanical forces [16,17,25]. As described in the anatomy section earlier, a combination of spine slenderness and thin or narrow physique is more common in girls than in boys. In spines of men and women who do not have scoliosis, there is a sexual dimorphism in pedicle height and vertebral body height [26,27], which may partially explain why girls are more susceptible to scoliosis forces but does not address why those forces begin.

Natural history

Curve progression

Guo and colleagues [25] postulated that after the scoliosis is established, the severity of the curve is proportional to the ratio of vertebral body height to pedicle height. This theory is supported by the finding that girls, who have higher proportional vertebral height to transverse diameter, are more likely to have progression of their curves.

Clinically, Suh and MacEwan [28] found that although curve progression secondary to growth typically stops when girls reach Risser stage 4, curves in scoliotic boys demonstrate significant progression until a boy reaches Risser stage 5. Thus, most investigators agree that boys who have curves over 20° to 30° should be followed radiographically until they reach Risser stage 5. Beyond Risser 5, male curves demonstrate minimal progression.

In his natural history study of scoliosis, Bunnell [1] found that curves in girls were more likely to progress if they were greater than 30°. In contrast, he found that patients of both sexes who had curves less than 30° had an equal risk of progression. Risk factors for curve progression in girls include sex, remaining skeletal growth, curve location, and curve magnitude [29–31]. In boys, however, Karol and colleagues [32] found that curve progression was related to immature Risser status but not to age or curve magnitude. These investigators found that progression to surgery was related to immature Risser status in addition to initial curve magnitude.

Treatment

Nonoperative treatment

Recent reports of controlled treatment trials using bracing have been very encouraging in girls [33], with curve progression at the end of treatment limited to less than 5° in 74% of patients treated with a brace compared with 34% in the group without treatment. In contrast, Karol [24] found that bracing in boys is ineffective. She found progression of 6° or greater in 74% of patients, particularly in younger patients who had larger curves, and progression to surgical magnitude in 46% despite the prescription of a brace. Karol [24] believed that low compliance was a factor in boys.

Operative treatment

Studies comparing surgical outcomes in boys and girls are limited. Helenius and colleagues [22] recently reported that posterior surgery for AIS provides similar short- and long-term results in boys and girls. Sucato and coworkers [17] noted that the curve magnitudes in male patients are typically greater than those of female patients at the time of surgery (62° ± 11° versus 56° ± 10°). These investigators found that male patients have longer operative times, greater blood loss, and less coronal plane correction of the primary curve. Satisfactory balance in the coronal and sagittal planes can be achieved, however, and complication rates and functional outcomes can be expected to be similar to those in female patients.

Summary

Sexual dimorphism in spine growth, morphology, stiffness, curve pattern, and hormones may be the environment in which genetic factors combine to produce the phenotype of a scoliosis patient. These factors also may play a role in curve progression despite treatment and may help explain why some patients' curves never change and others are recalcitrant to nonoperative treatments.

References

[1] Bunnell WP. The natural history of idiopathic scoliosis before skeletal maturity. Spine 1986;11:773–6.

[2] Lenke L, Dobbs M. Idiopathic scoliosis. In: Frymoyer J, Wiesel S, editors. The adult and pediatric spine, vol . 1. 3rd edition. Philadelphia: Lippincott, Williams & Wilkins; 2004. p. 337–60.

[3] Wenger D, Mubarak S, Chambers H, et al. Pediatric scoliosis. In: Garfin S, Vaccaro A, editors. Orthopaedic knowledge update: spine. Rosemont (IL): AAOS; 1997. p. 183–94.

[4] Riseborough EI, Wynne-Davies R. A genetic survey of idiopathic scoliosis in Boston, Massachusetts. J Bone Joint Surg Am 1973;55:974–82.

[5] Kesling K, Reinker K. Scoliosis in twins: a meta-analysis of the literature and report of 6 cases. Spine 1997;22:2009–14.

[6] Ahn UM, Ahn NU, Nallamshetty L, et al. The etiology of adolescent idiopathic scoliosis. Am J Orthop 2002;387–95.

[7] Garland HG. Hereditary scoliosis. Br Med Bull 1934;6:228–33.

[8] Miller NH. Cause and natural history of adolescent idiopathic scoliosis. Orthop Clin North Am 1999;30(3):343–52.

[9] Miller NH, Mims B, Child A, et al. Genetic analysis of structural elastic fiber and collagen genes in familial adolescent idiopathic scoliosis. J Orthop Res 1996;14:994–9.

[10] Miller NH, Sponseller PD, Bell J, et al. X chromosomes analysis in adolescent idiopathic scoliosis. Research into spinal deformities. 2. In: Grivas TB, editor. Proceedings of the 2nd Biannual Meeting of the International Research Society of Spinal Deformities. Burlington (VT): Amsterdam IOS Press; 1998.

[11] Robin JC, Cohen T. Familial scoliosis—a case report. J Bone Joint Surg 1975;57:146–7.

[12] Wynne-Davies R. Familial idiopathic scoliosis. J Bone Joint Surg Br 1968;50:24–30.

[13] Cowell HR, Hall IN, McEwen GD. Genetic aspects of idiopathic scoliosis. A Nicholas Andry Award essay 1972. Clin Orthop 1972;86:121–31.

[14] Justice C, Miller N, Marosy B, et al. Familial idiopathic scoliosis: evidence of an X-linked susceptibility locus. Spine 2003;15(28):589–94.

[15] Taylor JR, Twomey LT. A sexual dimorphism in human vertebral body shape. J Anat 1984;138:281–6.

[16] Burwell RG. The aetiology of idiopathic scoliosis: current concepts. Ped Rehab 2003;6:3–4, 137.

[17] Burwell RG. The relationship between scoliosis and growth. In: Zorab PA, editor. Scoliosis and growth. Proceedings of a third symposium. Edinburgh and London: Churchill Livingstone; 1971. p. 131–50.

[18] Sucato DJ, Hedequist D, Karol LA. Operative correction of adolescent idiopathic scoliosis in male patients. A radiographic and functional outcome comparison with female patients. J Bone Joint Surg Am 2004;86(9):2005–14.

[19] Goldberg CJ, Fogarty EE, Dowling FE. Left thoracic curve patterns and their association with disease. Spine 1998;24:1228–33.

[20] Grivas TB, Samblis P, Pappa AS, et al. Menarche in scoliotic and nonscoliotic Mediterranean girls. Is there a relation between menarche and laterality of scoliotic curves. In: Tangay, Peachot B, editors. Research into spinal deformities. Amsterdam, The Netherlands: IOS Press; 2002. p. 30–6.

[21] Dickson RA, Weinstein SL. Bracing (and screenings—yes or no?). J Bone Joint Surg Br 1999;81:193–8.

[22] Helenius I, Remes V, Yrjonen T, et al. Does gender affect outcome of surgery in adolescent idiopathic scoliosis? Spine 2005;30(4):462–7.

[23] Barrios C, Perez-Encivas C, Escriva D, et al. Body composition profile of girls with idiopathic scoliosis. Eur Spine J 12(Suppl 1):2.

[24] Karol LA. Effectiveness of bracing in male patients with idiopathic scoliosis. Spine 2001;26:2001–5.

[25] Guo X, Chau WW, Chan YL, et al. Relative anterior spinal overgrowth in adolescent idiopathic scoliosis. Results of disproportionate endochondral-membranous bone growth. J Bone Joint Surg Br 2003;85:1026–31.

[26] O'Higgins P, Johnson DR. The inheritance of vertebral shape in the mouse III. A study using Fourier

analysis to examine the inheritance of patterns of vertebral variation in the cervical and upper thoracic vertebral column. J Anat 1993;182:65–73.

[27] Schultz AB, Sorensen SE, Anderson GBJ. Measurements of spine morphology in children, ages 10–16. Spine 1984;9:70–3.

[28] Suh PB, MacEwen GD. Idiopathic scoliosis in males. A natural history study. Spine 1983;13:1091–5.

[29] Lonstein JE, Carlson JM. The prediction of curve progression in untreated idiopathic scoliosis during growth. J Bone Joint Surg Am 1984;66:1061–71.

[30] Weinstein SL, Zavala DC, Ponseti IV. Idiopathic scoliosis. Long-term follow-up and prognosis in untreated patients. J Bone Joint Surg Am 1981;63:702–12.

[31] Weinstein SL. The pediatric spine. 2nd edition. Philadelphia: Lippincott, Williams & Wilkins; 2001.

[32] Karol LA, Johnston CE, Browne RH, et al. Progression of the curve in boys who have idiopathic scoliosis. J Bone Joint Surg Am 1993;75:1804–10.

[33] Nachemsson AL, Peterson LE. Effectiveness of treatment with brace in girls who have adolescent idiopathic scoliosis: a prospective, controlled study based on data from the brace study of the Scoliosis Research Society. J Bone Joint Surg Am 1995;77:815–22.

Osteoarthritis of the Hip and Knee: Sex and Gender Differences

Mary I. O'Connor, MD

Department of Orthopedic Surgery, Mayo Clinic, 4500 San Pablo Road, Jacksonville, FL 32224, USA

Osteoarthritis (OA) of the hip and knee is a significant public health issue and a leading cause of functional disability and compromised quality of life in older patients [1]. Emerging research shows sex and gender differences in OA which, to date, may not be appreciated by the orthopedic community. This article discusses sex and gender differences in OA with a focus on disease involving the hip and knee. Understanding what we know (and do not know) about sex and gender differences in this disorder is critical to improving quality of care for our patients.

Epidemiology: how does the incidence of osteoarthritis differ between men and women?

OA is the most common chronic illness in the United States; it affects 59% of Americans 65 years of age or older. The overall prevalence of OA is higher in women as compared with men [2]. Reasons for this are not clear. Although genetics has been shown to influence OA, sex differences also may influence this prevalence.

Meta-analysis and meta-regression have been used to better define site-specific sex differences in prevalence, incidence, and severity of OA [3]. Table 1 details these findings. Men have a significantly reduced risk of knee and hand OA but a greater risk of cervical spine disc degeneration as compared with women. Women, especially those older than 55 years, had more severe OA in the knee but not other sites. More research is needed to better understand these epidemiology differences.

E-mail address: oconnor.mary@mayo.edu

Anatomy and physiology: how do they differ between men and women?

Given this sex difference in burden of disease, investigations have focused on the role of sex hormones on the development of OA. Estrogen receptors are found in many cell types, including human articular cartilage and bone [4–6]. Thus, human articular cartilage is hormonally sensitive.

Studies have focused on estrogen receptor genes. Investigating individuals in the Rotterdam Study (a large, population-based study of elderly white individuals living in the Netherlands), Bergink and colleagues [7] identified a specific allele (haplotype PX) of the estrogen receptor α gene that placed homozygous individuals at significant increased risk for knee OA. Risk estimates were noted, however, to be similar for men and women. Ushiyama and colleagues [6] also reported an increased risk for generalized OA in Japanese women with estrogen gene haplotypes.

Reviewing the data of Ushiyama and colleagues [6], Bergink and colleagues [7] noted a different distribution of the haplotype allele PX between the Dutch and Japanese patients. This suggests that differences in the incidence of OA between population groups may be influenced genetically. Bergink and colleagues also reviewed data on English patients [8] and found a nonsignificant increase in the frequency of the PX haplotype in men who had undergone total joint surgery, but a decrease in the frequency of the PX haplotype in women who had undergone joint replacement surgery. The investigators noted that the reason for this difference in association is unclear. More research is needed to understand better the influence of genetics and potential sex-based genetic differences on the development and progression of OA.

Table 1
Sex differences in prevalent osteoarthritis

Patient age	Any OA	Knee OA	Hip OA	Hand OA
Overall estimate	0.93	0.63	1.18	0.81
Age <55 y	0.98	0.82	1.04	1.03
Age >55 y	0.92	0.65[a]	1.05	0.77[a]

Numbers listed are relative risk and refer to the ratio of the prevalence of OA among men to the prevalence of OA among women. A ratio of less than 1 indicates a higher prevalence in women.

[a] These relative risks were statistically significant at $P = .05$.

Data from Srikanth VK, Fryer JL, Zhai G, et al. A meta-analysis of sex differences prevalence, incidence and severity of osteoarthritis. Osteoarthritis Cartilage 2005;13(9):769–81.

Differences in cartilage thickness also have been noted between genders. In a small study using quantitative three-dimensional MRI, Faber and colleagues [9] found cartilage thickness of the distal femur to be less in women (1.65–2.01 mm) than in men (1.73–2.05 mm). Whether this contributes to the higher incidence of OA in women has not been shown, but certainly, it could be a contributory factor.

Once OA is established, sex-related anatomic differences exist that are pertinent to surgical treatment. Generally, the female distal femur is narrower than the male femur for a given anterior–posterior dimension [10]. Hitt and colleagues [11] analyzed the exact anatomic morphologic data from the distal femur, proximal tibia, and patella from 337 knees undergoing total knee arthroplasty (TKA); they correlated these findings with sizing of knee implant systems available at that time. The investigators found that femoral components for women tended to be too large in the medial–lateral dimension for a given anterior–posterior dimension. In men, the femoral component tended to be smaller than the morphologic data in the medial–lateral dimension for a given anterior–posterior dimension Differences in size ratios (anterior–posterior dimension/medial–lateral dimension) were noted between implants of various manufacturers.

Size ratio mismatch of the femoral component to the patient's bone can have significant clinical implications. Use of a femoral component that is too large in the medial–lateral dimension will result in overhang of the component and may result in postoperative pain and dysfunction that are related to soft tissue irritation and ligament imbalance. If a smaller femoral component is used to avoid medial–lateral overhang, the component will be relatively smaller in the anterior–posterior dimension, which necessitates overresection of the posterior femoral condyles (Fig. 1) or notching of the anterior femoral cortex. Overresection of the posterior femoral condyles is associated with postoperative knee instability in flexion [12,13]; notching of the anterior cortex weakens the femur [14], which, in some patients, may increase the risk for postoperative distal femoral fracture.

Hitt and colleagues [11] also noted a difference in patellar height and tibial condylar ratios. The average unresected patella was 25.3 mm in men and 22.5 mm in women. This requires an implant system to have 8- to 10-mm thick patellar components to maintain 10 to 12 mm of remaining patellar bone for women. Moreover, the size of the lateral tibial condyle relative to the medial tibial condyle was smaller in women. This difference may be important, because in a woman a tibial implant that provides good coverage of the medial tibial condyle may be large for the lateral tibial condyle. This may cause popliteal tendon irritation and pain due to overhang of the component in the posterolateral corner of the knee (Fig. 2).

Finally, sex differences in the Q-angle have been documented. Women have about 3° greater Q-angle than do men [15]. This difference in Q-angle was confirmed to exist in the supine and standing positions [16].

Recently, some implant manufacturers have modified their femoral component sizing based on this anthropometric data (Gender Solutions, Zimmer, Warsaw, Indiana; Triathlon Knee System, Stryker, Mahwah, New Jersey) or designed femoral components specifically for women (Gender Solutions). Data indicating that such design changes will impact clinical outcomes are, to the author's knowledge, lacking. In one study of 268 patients who underwent bilateral TKA (68% women), component sizes were selected based on preoperative templating and intraoperative sizing measurements, regardless of the size that had been used in the contralateral knee arthroplasty [17]. Of the 269 bilateral TKAs, 18 (6.7%) varied in size of the femoral component between knees. The investigators found no statistical difference with respect to knee score, pain, function, range of motion, need for lateral retinacular release, or complications. Larger datasets are needed to study this sizing question, and subtle functional and clinical issues that are related to sizing (eg, anterior knee pain, flexion stability) need to be specifically investigated.

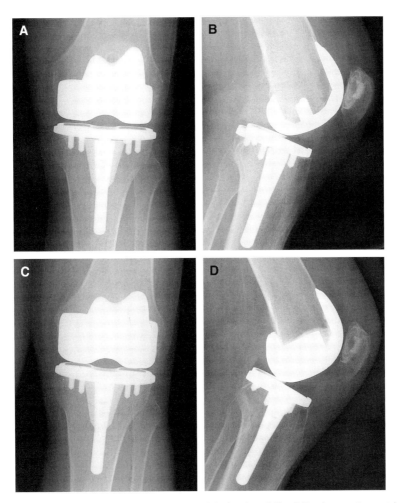

Fig. 1. Seventy-year-old patient with marked knee pain and flexion instability following cruciate retaining TKA. (A) Anteroposterior view shows that femoral component size seems to be satisfactory. (B) Lateral film, however, shows marked overresection of the posterior femoral condyles. (C, D) Revision surgery was performed successfully to a larger femoral component with posterior condylar augments. The original polyethylene insert was a flat design; the revision insert was a posterior substituted design with extended post to provide additional knee stability.

Risk factors and prevention: do they differ between men and women?

Various factors may influence the risk for development of OA. Generally, accepted risk factors include advancing age, hip dysplasia, obesity, and previous joint injury. Other factors that may influence OA development are use of postmenopausal estrogen, level of exercise, and osteoporosis (OP). Some of these factors clearly differ between men and women.

Osteoporosis and osteoarthritis

The relationship between OP and OA has been studied by many investigators. Due to early data noting infrequent OA in patients who had osteoporotic hip fractures (Fig. 3) [18] and subsequent bone mineral density (BMD) studies, which, in general, showed that patients who had OP had a lower than expected incidence of OA [19], many believed that OP had a protective effect relative to OA. Hypotheses for this association included that greater bone mass increased subchondral bone stiffness, and, therefore, increased joint loading and the risk for joint OA.

More recent studies continue to investigate this relationship. In support of a protective effect of OP relative to OA, Sowers and colleagues [20] found that premenopausal women who had knee OA had greater BMD than did those who did

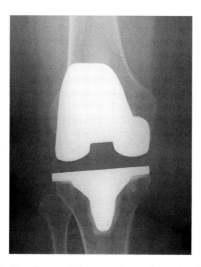

Fig. 2. Standing posterior–anterior radiograph of a 58-year-old woman who had undergone TKA. She had postoperative complaints of pain in the posterolateral knee and tenderness in this region on examination. Radiograph shows slight lateral overhang of the tibial component. Clinical and radiographic findings are consistent with pain due to popliteal tendon irritation. Arthroscopic release of the popliteus tendon was performed with resolution of discomfort.

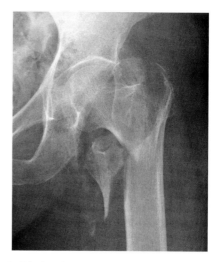

Fig. 3. Displaced osteoporotic-related intertrochanteric hip fracture in an elderly white woman. Note good preservation of hip joint space. Patient had no hip pain before fracture.

not have knee OA and were less likely to lose that higher level of BMD. Levels of osteocalcin also were lower in women who had knee and hand OA, which suggested less bone turnover in the individuals who had OA. In a study of forearm BMD in postmenopausal women, Iwamoto and colleagues [21] also found significantly higher BMD in women who had knee OA as compared with controls. Of interest was the finding that BMD increase was significantly higher in individuals who had Kellgren and Lawrence radiologic grades 2, 3, and 4 OA compared with grade 1 OA. Furthermore, BMD was higher in women who had grade 3 OA as compared with grade 2 OA; however, BMD in women who had grade 4 OA was significantly lower than in those who had grade 3 changes. The investigators concluded that some cases of severe OA may be associated with low BMD. This is consistent with evidence reported by Terauchi and colleagues [22] that varus OA of the knee may be associated with high BMD or low BMD, in which trabecular microfracture of the proximal tibia may create the varus deformity with secondary OA. Moreover, in a study of postmenopausal white women who had severe OA and underwent total hip arthroplasty, Glowacki and colleagues [23] found that 25% had occult OP and 22% had vitamin D deficiency. Vitamin D deficiency was not restricted to those women who had low BMD. As concluded by Glowacki and colleagues, the diagnosis of OA does not eliminate the risk for accelerated bone loss, and all postmenopausal women should be evaluated for BMD.

Longitudinal data on the influence of BMD and OA in women are available from the Framingham population (elderly white postmenopausal women) [24]. Over 8 years of follow-up, radiographic knee OA increased in women with higher BMD, which primarily reflected an increased risk for osteophyte development. Once OA was established, however, women with higher BMD and BMD gain showed less progression of disease (less joint space narrowing). The investigators hypothesized that bone effects may differ at different states of OA.

Developmental dysplasia of the hip

Developmental dislocation (or dysplasia) of the hip has long been recognized as a risk factor for the development of hip OA and is predominant in women in the United States. It has been estimated that 20% to 40% of patients who undergo hip arthroplasty for OA have some degree of underlying developmental dysplasia of the hip [25,26]. Patients who have severe dysplasia and hip subluxation are known to be at predictable risk for premature OA [27,28]; however, individuals who have mild hip dysplasia have an unpredictable course in terms of onset and progression of hip OA [29].

Are women who have mild acetabular dysplasia at a higher risk for developing hip OA than are men who have mild dysplasia? The answer is unclear. In a prospective cohort design with 8 years of follow-up, acetabular dysplasia (center-edge angle of $<30°$) was associated with only a modestly increased risk for incident hip OA in elderly white women [30]. In the Rotterdam study of adults aged 55 years or older who had no radiographic signs of hip OA at baseline, acetabular dysplasia was a strong determinant of incident hip OA at mean follow-up of 6.6 years. Although women who had acetabular dysplasia developed joint space narrowing more often during the follow-up period, the association between dysplasia and OA was independent of age, sex, body mass index (BMI), and follow-up time [31]. In fact, the Rotterdam study noted that the prevalence of acetabular dysplasia was similar in men and women. Likewise, the Copenhagen Osteoarthritis Study cohort noted men and women to have similar acetabular dysplasia, with the risk for hip OA significantly influenced by hip dysplasia in men and by hip dysplasia and age in women [32]. The Copenhagen study found that the risk for hip arthroplasty being performed was influenced only by the patient's BMI at the onset of the study and was not related to dysplasia, however. Also of interest is a small study of Turkish men and women in which acetabular dysplasia was more common in men (13%) than in women (3.7%), but it was not identified as a significant factor in the development of hip OA [33]. Certainly, population differences may exist with hip dysplasia and the development of hip OA. Moreover, risk factors for surgical intervention may differ to some degree from risk factors for development of OA.

Postmenopausal estrogen use

The role of postmenopausal estrogen use in the risk for the development or progression of OA is unclear. In white women in the United Kingdom, postmenopausal estrogen had a nonsignificant protective effect on the risk for developing OA [34]. Data from a longitudinal study of white women in southern California, however, showed that women who used postmenopausal estrogen were more likely to have hip and hand OA than were controls, even with adjustment for age, BMI, smoking, exercise, and onset of menopause [35].

Influence of exercise

The influence of exercise on hip OA in women was studied by Lane and colleagues [36]. In a study of participants in the Study of Osteoporotic Fractures (white women 65 years of age or older from four areas of the United States), the investigators found that recreational physical activities before menopause may increase the risk for radiographic and symptomatic hip OA. In a subsequent study, Lane and colleagues [37] noted that women in this study who had mild hip joint space narrowing (1.5–2.5 mm) were unlikely to show radiographic progression of disease over 8 years. In men, some investigators found a positive association between high and moderate levels of physical activity and hip OA [38,39], whereas others reported that former elite and recreational runners aged 50 to 65 years had no increased risk for hip OA [40,41].

The relationship between exercise and knee OA also has been studied with similar conflicting results. Spector and colleagues [42] reported higher rates of radiographic hip and knee OA in former elite female runners and tennis players. Studying recreational athletes from the Framingham cohort, Hannan and colleagues [43] found no association between recreational, habitual exercise in middle age and the development of knee OA in later years, after adjusting for confounding variables. White and colleagues [44] reported a lower prevalence of knee OA and a similar prevalence of hip OA in middle-aged women who were physical education teachers compared with a closely age-matched group; at 12 years of follow-up these women reported less joint stiffness and pain.

Although clear conclusions cannot be drawn based on these data, high-intensity athletes (male and female) may be at higher risk for hip and knee OA because of their type and intensity of exercise. Individuals who engage in lower-level recreational activities may or may not be at increased risk for hip or knee OA. Some of these studies are limited by recall of the individuals in terms of their previous level of exercise. Genetic factors also were not analyzed and certainly could influence the results.

Body weight

Heavy weight was shown to increase the risk for knee OA in American, British, and Japanese populations [34,45–48]. Obesity influenced the risk for hip arthroplasty in the Copenhagen study [32].

Perhaps of greater importance, weight reduction can reduce the risk for developing symptomatic knee OA in women [49]. The critical importance of appropriate body weight needs to be emphasized to all patients who have knee OA.

In patients who undergo joint replacement surgery, obesity has not been identified clearly as a negative factor in overall outcomes. The Agency for Heath Care Research and Quality (AHRQ) evidence report on TKA [50] found no evidence that obesity is a strong predictor of functional outcome following knee replacement surgery. In a study of hip and knee arthroplasty, obesity did not affect outcome adversely [51].

Prior knee injury

Long-term consequences of anterior cruciate ligament (ACL) injury among female Swedish soccer players are troubling [52]. Eighty-four women (ages 26–40 years) who had sustained an ACL injury 12 years earlier were studied. At the time of the study, 75% had symptoms that substantially affected their quality of life and 42% had symptomatic radiographic knee OA. No difference was noted in women who had undergone ACL reconstruction compared with those whose injury was managed nonoperatively. Although the risk for knee OA after ACL injury is well established regardless of sex, women have a higher risk for ACL injury. In the near future, American orthopedic surgeons may see an increase in young women who have significant knee OA that is related to such previous sports injuries. Treatment of such a young patient cohort will, most likely, prove challenging. Efforts to prevent ACL injuries in athletes, particularly female, must be increased.

Diagnosis: do men and women present the same or differently?

The diagnosis of OA of the hip or knee does not differ between genders. Patients who have severe hip or knee OA typically present with significant joint pain, painful and limited ambulation, discomfort that is exacerbated with weight-bearing activities, and marked joint space narrowing and osteophyte formation.

Men and women may present differently in terms of severity of symptoms, however. Studies have suggested a discrepancy in health care interventions between genders. Investigating differences between men and women in the rate of use of hip and knee arthroplasty, Hawker and colleagues [53] found that Canadian women had a higher prevalence of hip and knee arthritis, worse symptoms, and greater disability; however, these same women were less likely to have undergone arthroplasty. Although hip and knee replacement surgery is performed more frequently in women than in men in the United States and Canada [54,55], such rates are not adjusted relative to the higher incidence of OA in women. Hawker and colleagues [53] reported the number of people with a potential need for hip or knee arthroplasty to be 44.9 per 1000 among women and 20.8 per 1000 among men. After adjusting for patient willingness to undergo arthroplasty, 5.3 per 1000 women and 1.6 per 1000 men were surgical candidates. The investigators concluded that joint replacement surgery for severe hip and knee OA is underused by both genders, but the degree of underuse is three times greater in women. Of note is that studies of other types of surgical procedures, such as coronary artery bypass grafting and renal transplantation, suggest that these procedures also are performed less often in women than they should be [56–61]. Thus, it seems that such gender differences in health care interventions involve many disciplines and are not isolated to orthopedic procedures.

Underuse of arthroplasty probably has many contributing factors. The National Institutes of Health (NIH) consensus conference [62] on TKA detailed the process of evaluation for knee arthroplasty. The conference noted that to receive a knee replacement, the patient must seek a physician for evaluation and then be referred to an orthopedic surgeon, or self-refer directly. The TKA procedure must be offered by the surgeon and the patient must be willing to proceed. Gender (and racial) differences in the use of TKA can result from inequities in any portion of this process.

Although reasons for presentation with greater symptoms and disability by women are not clear, a most likely explanation is that women present to orthopedic surgeons later in the course of their disease. Factors that may contribute to this difference include lack of social support at home and the patient's willingness (or lack thereof) to undergo surgical treatment. In an effort to understand this disparity, Chang and colleagues [63] recruited 37 patients for focus group study (12 men, 25 women; 20 self-identified as white and 17 as African American). White women was the only group to ask questions regarding potential drawbacks to surgery; African American women wanted to understand the criteria for TKA and their individual needs for arthroplasty; white men had the greatest background knowledge regarding TKA and expressed interest in implants and technology; and African American men noted concerns regarding health insurance coverage of the surgery. Women asked the most questions about preoperative and postoperative care, whereas men raised fewer issues than did the women. Moreover, white men asked questions that were answered more easily by surgeons ("What happens to the patella in TKA?"). Apparently, this tendency for white men to speak in a language that is more consistent with that of their physicians has been documented by medical anthropologists in studying doctor–patient relationships. Chang and colleagues [63] noted that this communications difference and the greater background knowledge could contribute to the higher use of TKA by white men. Karlson and colleagues [64] also found that women seemed to have less background information about TKA than did men. Certainly, further study of gender and racial differences is needed to allow orthopedic surgeons to educate and communicate with patients more effectively.

Are treatment methods different between men and women?

Joint replacement surgery is the treatment for severe, disabling hip and knee OA. There is consensus that such procedures are effective and consistently relieve pain and improve function. In a study of preoperative functional status before and 6 months following total hip or knee arthroplasty, patients with lower preoperative physical function did not improve following surgery to the same degree as did patients with less impairment before surgery [65]. The investigators noted this was most striking in patients who underwent total knee replacement. Other investigators also have noted that patients with more severe pain, greater functional limitations, low mental health scores, and other comorbid conditions before surgery are more likely to have a worse outcome [66,67].

With data suggesting that women undergo surgery later in the course of their arthritic disease process, concern has been raised regarding sex differences in outcomes. Two different Canadian studies looked at whether a longer wait for total hip replacement affected outcome. Mahon and colleagues [68] prospectively studied 99 patients who underwent total hip arthroplasty. They found no significant difference in postoperative health-related quality of life or mobility in patients who waited less than or more than 6 months for surgery; however, at the time of surgical referral the patients with shorter wait times had poorer health-related quality of life and mobility than did those who waited longer than 6 months. Garbuz and colleagues [69] prospectively examined 201 patients who had hip OA and were awaiting arthroplasty. Logistic regression models were used to assess the relationship between waiting time and the probability of a better than expected outcome. The investigators found that the odds of achieving a better than expected functional outcome after surgery decreased by 8% for each month on the waiting list. Although delay in surgery may be a negative factor in postoperative outcomes, regardless of sex, the NIH consensus panel [62] and the AHRQ evidence-based report on TKA [50] did not find sex to be a strong predictor of functional outcome following TKA, based on current measurement tools.

Summary

Sex and gender differences exist in OA. Our understanding of these differences is at a preliminary stage. Further research, specifically to address sex and gender differences, is needed. With the current knowledge, an effort should be made to decrease the burden of disease that is related to OA. These efforts should focus on improved education of patients, particularly women, to decrease the risk for OA and progression of degenerative changes, as well as optimal outcomes following joint replacement surgery.

References

[1] Hootman JM, Sniezek JE, Helmick CG. Women and arthritis: burden, impact and prevention programs. J Womens Health Gend Based Med 2002; 11(5):407–16.

[2] Peyron JG, Altman RD. The epidemiology of osteoarthritis. In: Moskowitz RW, Howell DS, Goldberg VM, et al, editors. Osteoarthritis: diagnosis and medical/surgical management. 2nd edition. Philadelphia: W.B. Saunders; 1992. p. 15–37.

[3] Srikanth VK, Fryer JL, Zhai G, et al. A meta-analysis of sex differences prevalence, incidence and severity of osteoarthritis. Osteoarthritis Cartilage 2005;13(9):769–81.

[4] Richmond RS, Carlson CS, Register TC, et al. Functional estrogen receptors in adult articular cartilage: estrogen replacement therapy increases chondrocyte

[5] Tsai CL, Liu TK, Chen TJ. Estrogen and osteoarthritis: a study of synovial estradiol and estradiol receptor binding in human osteoarthritic knees. Biochem Biophys Res Commun 1992;183(3):1287–91.

[6] Ushiyama T, Ueyama H, Inoue K, et al. Expression of genes for estrogen receptors alpha and beta in human articular chondrocytes. Osteoarthritis Cartilage 1999;7(6):560–6.

[7] Bergink AP, van Meurs JB, Loughlin J, et al. Estrogen receptor alpha halotype is associated with radiographic osteoarthritis of the knee in elderly men and women. Arthritis Rheum 2003;48(7):1913–22.

[8] Loughlin J, Sinsheimer JS, Mustafa Z, et al. Association analysis of the vitamin D receptor gene, the type I collagen gene COL1A1, and the estrogen receptor gene in idiopathic osteoarthritis. J Rheumatol 2000;27(3):779–84.

[9] Faber SC, Eckstein F, Lukasz S, et al. Gender differences in knee joint cartilage thickness, volume and articular surface areas: assessment with quantitative three-dimensional MR imaging. Skeletal Radiol 2001;30(3):144–50.

[10] Poilvache PL, Insall JN, Scuderi GR, et al. Rotational landmarks and sizing of the distal femur in total knee arthroplasty. Clin Orthop Relat Res 1996;331:35–46.

[11] Hitt K, Shurman JR II, Greene K, et al. Anthropometric measurements of the human knee: correlation to the sizing of current knee arthroplasty systems. J Bone Joint Surg Am 2003;85-A (Suppl 4):115–22.

[12] Clarke HD, Scuderi GR. Flexion instability in primary total knee replacement. J Knee Surg 2003;16(2):123–8.

[13] Schwab JH, Haidukewych GJ, Hanssen AD, et al. Flexion instability without dislocation after posterior stabilized total knees. Clin Orthop Relat Res 2005;440:96–100.

[14] Lesh ML, Schneider DJ, Deol G, et al. The consequences of anterior femoral notching in total knee arthroplasty. A biomechanical study. J Bone Joint Surg Am 2000;82-A(8):1096–101.

[15] Hsu RW, Himeno S, Coventry MB, et al. Normal axial alignment of the lower extremity and load-bearing distribution at the knee. Clin Orthop Relat Res 1990;255:215–27.

[16] Woodland LH, Francis RS. Parameters and comparisons of the quadriceps angle of college-aged men and women in the supine and standing positions. Am J Sports Med 1992;20(2):208–11.

[17] Brown TE, Diduch DR, Moskal JT. Component size asymmetry in bilateral total knee arthroplasty. Am J Knee Surg 2001;14(2):81–4.

[18] Foss MV, Byers PD. Bone density, osteoarthrosis of the hip, and fracture of the upper end of the femur. Ann Rheum Dis 1972;31(4):259–64.

[19] Dequeker J, Aerssens J, Luyten FP. Osteoarthritis and osteoporosis: clinical and research evidence of inverse relationship. Aging Clin Exp Res 2003;15(5):426–39.

[20] Sowers M, Lachance L, Jamada D, et al. The association of bone mineral density and bone turnover markers with osteoarthritis of the hand and knee in pre- and postmenopausal women. Arthritis Rheum 1999;42(3):483–9.

[21] Iwamoto J, Takeda T, Ishimura S. Forearm bone mineral density in postmenopausal women with osteoarthritis of the knee. J Orthop Sci 2002;7(1):19–25.

[22] Terauchi M, Shirakura K, Katayama M, et al. The influence of osteoporosis on varus osteoarthritis of the knee. J Bone Joint Surg Br 1998;80(3):432–6.

[23] Glowacki J, Hurwitz S, Thornhill TS, et al. Osteoporosis and vitamin-D deficiency among postmenopausal women with osteoarthritis undergoing total hip arthroplasty. J Bone J Surg Am. 2003 Dec;85-A(12):2371–7.

[24] Zhang Y, Hannan MT, Chaisson CE, et al. Bone mineral density and risk of incident and progressive radiographic knee osteoarthritis in women: the Framingham Study. J Rheumatol 2000;27(4):1032–7.

[25] Harris WH. Etiology of osteoarthritis of the hip. Clin Orthop Relat Res 1986;213:20–33.

[26] Solomon L. Patterns of osteoarthritis of the hip. J Bone Joint Surg Br 1976;58(2):176–83.

[27] Wedge JH, Wasylenko MJ. The natural history of congenital dislocation of the hip: a critical review. Clin Orthop Relat Res 1978;137:154–62.

[28] Weinstein SL. Natural history of congenital hip dislocation (CDH) and hip dysplasia. Clin Orthop Relat Res 1987;225:62–76.

[29] Cooperman DR, Wallensten R, Stulberg SD. Acetabular dysplasia in the adult. Clin Orthop Relat Res 1983;175:79–85.

[30] Lane NE, Lin P, Christiansen L, et al. Association of mild acetabular dysplasia with an increased risk of incident hip osteoarthritis in elderly white women: the study of osteoporotic fractures. Arthritis Rheum 2000;43(2):400–4.

[31] Reijman M, Hazes JM, Pols HA, et al. Acetabular dysplasia predicts incident osteoarthritis of the hip: the Rotterdam study. Arthritis Rheum 2005;52(3):787–93.

[32] Jacobsen S, Sonne-Holm S. Increased body mass index is a predisposition for treatment by total hip replacement. Int Orthop 2005;29(4):229–34.

[33] Goker B, Sancak A, Haznedaroglu S. Radiographic hip osteoarthritis and acetabular dysplasia in Turkish men and women. Rheumatol Int 2005;25(6):419–22.

[34] Hart DJ, Doyle DV, Spector TD. Incidence and risk factors for radiographic knee osteoarthritis in middle-aged women: the Chingford Study. Arthritis Rheum 1999;42(1):17–24.

[35] Von Munien D, Morton D, Von Muhlen CA, et al. Postmenopausal estrogen and increased risk of clinical osteoarthritis at the hip, hand, and knee in older women. J Womens Health Gend Based Med 2002; 11(6):511–8.

[36] Lane NE, Hochberg MC, Pressman A, et al. Recreational physical activity and the risk of osteoarthritis of the hip in elderly women. J Rheumatol 1999;26(4): 849–54.

[37] Lane NE, Nevitt MC, Hochberg MC, et al. Progression of radiographic hip osteoarthritis over eight years in a community sample of elderly white women. Arthritis Rheum 2004;50(5):1477–86.

[38] Marti B, Knobloch M, Tschopp A, et al. Is excessive running predictive of degenerative hip disease? Controlled study of former elite athletes. BMJ 1989; 299(6691):91–3.

[39] Vingard E, Alfredsson L, Goldie I, et al. Sports and osteoarthrosis of the hip. An epidemiologic study. Am J Sports Med 1993;21(2):195–200.

[40] Panush RS, Hanson CS, Caldwell JR, et al. Is running associated with osteoarthritis? An eight year follow-up study. J Clin Rheumatol 1995;1:35–9.

[41] Puramen J, Ala Ketola L, Peltokullio P. Running and osteoarthritis of the hip. BMJ 1975;2:424–5.

[42] Spector TD, Harris PA, Hart DJ, et al. Risk of osteoarthritis associated with long-term weight-bearing sports: a radiologic survey of the hips and knees in female ex-athletes and population controls. Arthritis Rheum 1996;39(6):988–95.

[43] Hannan MT, Felson DT, Anderson JJ, et al. Habitual physical activity is not associated with knee osteoarthritis: the Framingham Study. J Rheumatol 1993;20(4):704–9.

[44] White JA, Wright V, Hudson AM. Relationship between physical activity and osteoarthritis in ageing women. Public Health 1993 Nov;107(6):459–70.

[45] Felson DT. The epidemiology of knee osteoarthritis: results from the Framingham Osteoarthritis Study. Semin Arthritis Rheum 1990;20(3)(Suppl 1):42–50.

[46] Hochberg MC, Lethbridge-Cejku M, Scott WW Jr, et al. The association of body weight, body fatness and body fat distribution with osteoarthritis of the knee: data from the Baltimore Longitudinal Study of Aging. J Rheumatol 1995;22(3):488–93.

[47] Spector TD. The fat on the joint: osteoarthritis and obesity. J Rheumatol 1990;17(3):283–4.

[48] Yoshimura N, Nishioka S, Kinoshita H, et al. Risk factors for knee osteoarthritis in Japanese women: heavy weight, previous joint injuries, and occupational activities. J Rheumatol 2004;31(1):157–62.

[49] Felson DT, Zhang Y, Anthony JM, et al. Weight loss reduces the risk for symptomatic knee osteoarthritis in women. The Framingham Study. Ann Intern Med 1992;116(7):535–9.

[50] Kane RL, Saleh KJ, Wilt TJ, et al. Total knee replacement. Evidence Report/Technology Assessment No. 86 (Prepared by the Minnesota Evidence-based practice Center, Minneapolis, MN). Rockville, MD: Agency for Healthcare Research and Quality; December 2003. AHRQ Publication No. 04–E006–2.

[51] Stickles B, Phillips L, Brox WT, et al. Defining the relationship between obesity and total joint arthroplasty. Obes Res 2001;9(3):219–23.

[52] Lohmander LS, Ostenberg A, Englund M, et al. High prevalence of knee osteoarthritis, pain, and functional limitations in female soccer players twelve years after anterior cruciate ligament injury. Arthritis Rheum 2004;50(10):3145–52.

[53] Hawker GA, Wright JG, Coyte PC, et al. Differences between men and women in the rate of use of hip and knee arthroplasty. N Engl J Med 2000;342(14): 1016–22.

[54] Naylor CD, Anderson GM, Goel V, editors. Variation in surgical services over time and by site of residence. In: Patterns of health care in Ontario-an ICES practice atlas, 1st edition. Ottawa (Canada): National Printers; 1994.

[55] DeBoer D, Williams JI. Surgical services for total hip and total knee replacements: trends in hospital volumes and length of stay for total hip and total knee replacement. In: Badley EM, Williams JI, editors. Patterns of health care in Ontario: arthritis and related conditions. Toronto (Canada): Continental Press; 1998. p. 121–4.

[56] Bell MR, Berger PB, Holmes DR Jr, et al. Referral for coronary artery revascularization procedures after diagnostic coronary angiography: evidence for gender bias? J Am Coll Cardiol 1995;25(7):1650–5.

[57] Bergelson BA, Tommaso CL. Gender differences in percutaneous interventional therapy of coronary artery disease. Cathet Cardiovasc Diagn 1996;37(1): 1–4.

[58] Bloembergen WE, Mauger EA, Wolfe RA, et al. Association of gender and access to cadaveric renal transplantation. Am J Kidney Dis 1997;30(6):733–8.

[59] Giacomini MK. Gender and ethnic differences in hospital-based procedure utilization in California. Arch Intern Med 1996;156(11):1217–24.

[60] Gijsbers van Wijk CM, van Vliet KP, Kolk AM. Gender perspectives and quality of care: towards appropriate and adequate health care for women. Soc Sci Med 1996;43(5):707–20.

[61] Weitzman S, Cooper L, Chambless L, et al. Gender, racial, and geographic differences in the performance of cardiac diagnostic and therapeutic procedures for hospitalized acute myocardial infarction in four states. Am J Cardiol 1997;79(6):722–6.

[62] Consensus NIH. Statement on total knee replacement. Dec. 8–10, 2003. Available at: http://consensus.nih.gov/cons/117/117cdc_statementFINAL.html.

[63] Chang HJ, Mehta PS, Rosenberg A, et al. Concerns of patients actively contemplating total knee replacement: differences by race and gender. Arthritis Rheum 2004;51(1):117–23.

[64] Karlson EW, Daltroy LH, Liang MH, et al. Gender differences in patient preferences may underlie

differential utilization of elective surgery. Am J Med 1997;102(6):524–30.

[65] Fortin PR, Clarke AE, Joseph L, et al. Outcomes of total hip and knee replacement: preoperative functional status predicts outcomes at six months after surgery. Arthritis Rheum 1999;42(8):1722–8.

[66] Fitzgerald JD, Orav EJ, Lee TH, et al. Patient quality of life during the 12 months following joint replacement surgery. Arthritis Rheum 2004;51(1):100–9.

[67] Lingard EA, Katz JN, Wright EA, et al. Predicting the outcome of total knee arthroplasty. J Bone Joint Surg Am 2004;86-A(10):2179–86.

[68] Mahon JL, Bourne RB, Rorabeck CH, et al. Health-related quality of life and mobility of patients awaiting elective total hip arthroplasty: a prospective study. CMAJ 2002;167(10):1115–21.

[69] Garbuz DS, Xu M, Duncan CP, et al. Delays worsen quality of life outcome of primary total hip arthroplasty. Clin Orthop Relat Res 2006;447:79–84.

Sexual Dimorphism of the Foot and Ankle

Kathryn O'Connor, PT, Gwynne Bragdon, MD,
Judith F. Baumhauer, MD*

*Division of Foot and Ankle Surgery, University of Rochester School of Medicine,
601 Elmwood Avenue, Rochester, NY 14642, USA*

Lower extremity musculoskeletal injuries are extremely common. Sports-related sex differences, in addition to osteoporosis issues, have raised the level of social awareness that women's health care issues may be different than those of their male counterparts. Traditional research investigation for the foot and ankle is focused on shoe style differences and the effect that these shoes have had on foot pain and injury (eg, bunion, lesser toe malalignment). In addition to the extrinsic factor of footwear, there are intrinsic factors, such as foot structure, ligamentous laxity, muscle strength, and proprioception, which predispose individuals to injury. This article reviews the literature to examine the intrinsic and extrinsic differences between men and women in relationship to the foot and ankle and explores, where available, the influence that these foot and ankle factors have on injury.

Foot osteology and foot shape

Traditionally, women's sports shoes have been fabricated using a proportionally scaled down model of men's feet. Despite the increasing interest in sex-specific differences in injury, little attention has been given to the sexual dimorphism of the foot in regards to osteology and shape. Wunderlich and Cavanagh [1] examined anthropometric and demographic data of approximately 300 men and 500 women in the US Army. After discriminate analysis, they concluded that female feet and legs are not simply scaled down versions of the male counterparts. Particularly, the shape of the arch, lateral side of the foot, great toe, and ball of the foot, after normalized to foot length, demonstrated the greatest differences. The woman's foot was wider in the forefoot, had shorter arch length, and the metatarsals were shorter than a man's foot. They concluded that these differences should be incorporated into the design and manufacturing of women's sports shoes. In a separate study, Fessler and colleagues [2] also found sexual dimorphism in foot length proportional to stature (Table 1). Male feet have higher foot length/body height ratios than do female feet, and this extends across a variety of ethnic populations. Body weight, however, is not related clearly to differences in foot shape. Ashizawa and colleagues [3] found no sexual dimorphism between body mass index and foot shape. With the exception of these few studies, anatomic comparisons of the foot between men and women have been limited to injury incidence and foot measurement correlations, not gender differences of osteology and foot shape.

Anthropologic studies have been performed modeling male and female ankle and foot bones to allow for forensic identification of sex [4,5]. Measurements made of foot bones, specifically metatarsal, phalangeal, calcaneal, and talar models, have high correlations for discriminate analysis for sex, which further defines sexual dimorphism in the foot and ankle.

Although bone morphology for the foot and ankle are not available, morphology of the distal tibia is different between men and women when adjusted for height and weight and assessed by dual energy x-ray absorbometry scanning and

* Corresponding author.
  E-mail address: judy_baumhauer@urmc.rochester.edu (J.F. Baumhauer).

Table 1
Summary of some published findings on foot length as a proportion of stature for men and women

Citation	Population	n	Male foot length as a % of stature	Female foot length as a % of stature
Hrdlička 1935	Apache	83	14.93	14.58
Hrdlička 1935	Aztec	84	15.38	14.98
Hrdlička 1935	Cora	61	15.21	15.08
Hrdlička 1935	Maricopa	70	15.19	15.20
Hrdlička 1935	Mohave	71	15.15	14.62
Hrdlička 1935	Navaho	79	14.66	14.66
Hrdlička 1935	Oromi	75	15.45	13.73
Hrdlička 1935	Papago	80	15.07	14.83
Hrdlička 1935	Pima	83	14.99	14.83
Hrdlička 1935	Pueblos	183	14.88	14.68
Hrdlička 1935	Tarahumare	32	14.82	14.41
Hrdlička 1935	Tarasco	80	15.18	14.85
Hrdlička 1935	Southern Ute	70	15.12	15.16
Hrdlička 1935	Yuma	34	15.01	14.85
Hrdlička 1928	African–Americans	26	15.89	16.11
Hrdlička 1935	Caucasian–Americans	455	14.97	14.42
Hrdlička 1925 (after Quetelet, 1835?)	Belgians	?	15.8	15.0
Hrdlička 1925 (after Martin 1958?)	Lithuanian	?	14.6	14.4
Hrdlička 1925 (after Martin 1958?)	Letts (Latvian)	?	14.6	14.8
Hrdlička 1925 (after Martin 1958?)	Badenese (German)	?	16.0	15.5
Davis 1990 (after Todd & Lindala 1928)	African–Americans	135	14.7	14.6
Davis 1990 (after Todd & Lindala 1928)	Caucasian–Americans	136	14.3	13.5
Davis 1990 (after Davenport & Steggerda 1929)	African–Jamaicans	100	15.4	15.2
Davis 1990 (after Steggerda 1932)	Dutch	130	15.3	14.9
Robbins 1986	US age 14 and up	527	right 15.128 left 15.199	right 14.726 left 14.750
Anderson et al. 1956	US 18-year-olds	20	14.9	15.1
Davis 1990	African–American 18-26-year-olds	110	right 15.58 left 15.61	right 15.30 left 15.31
Davis 1990	Caucasian–American 18-26-year-olds	130	right 15.27 left 15.24	right 14.58 left 14.61
Giles and Vallandigham 1991	US soldiers	8012	15.346	14.926
Wunderlicb and Cavanagh 2001	US soldiers	754	15.36	15.01
Barker and Scheuer 1998	Predominantly Caucasian (London)	105	right footprint: 15.222 left footprint: 15.189	right footprint: 14.853 left footprint: 14.806

Data from Fessler DMT, Haley JK, Roshni DL. Sexual dimorphism in foot length proportionate to stature. Ann Hum Biol 2005;32(1):46.

histologic sectioning [3]. Female distal third tibial bones, on the average, were narrower and had thinner cortices than did the male bones.

Cartilage

Osteoarthritis of the ankle was reported to be more common in men [7]. Eckstein and colleagues [8] examined cartilage morphology of the knee, ankle, and foot by MRI to examine sexual differences in the lower leg. These cartilage values included volume, joint surface area, and cartilage thickness of the knee, patella, subtalar, talonavicular, and ankle joints. Although the primary purpose of the study was not to identify sexual differences in the ankle and foot specifically, there were statistically significant differences in the joint surface area and cartilage volume in the ankle, subtalar, and talonavicular joints. In each case, women had 20% to 25% lower surface area and volume in these joints, and their cartilage thickness was 11% to 16% thinner. The investigators had no comment on how these smaller values influenced degenerative disease or injury.

Muscle

Several studies have examined sexual differences in muscle strength, endurance, and torque of the foot and ankle. Although men have a greater mean torque of all major muscle groups in the lower extremity (women < 62%–70% of men), when normalized for size by bone mass index or body weight, there were no significant differences in strength, endurance, or torque [9]. Age-adjusted data, however, do demonstrate a loss of strength of great toe flexors for men and women. Women proportionately lose more strength than do men in the great toe as they age [10].

Gait

Range of motion, cadence, and walking speed were examined in two studies to identify gender differences and establish normative data in young adults [11,12]. A common isolated statistically significantly finding involved ankle plantar flexion. Cho and colleagues [11] found that women had a lower plantar flexion moment during stance phase. Kerrigan and colleagues [12] reported greater plantar flexion range of motion and greater ankle power generated in women during preswing phase of gait. Sepic and colleagues [9] also found greater ankle plantar flexion range of motion in women.

Sex differences in plantar pressure and contact area of the foot were examined by a Pedar pressure insole measurement device, and male and female data were compared [13]. Men had larger normalized contact area and plantar pressure values during midstance than did women; however, this was not statistically significant. There were no other statistically significant differences in plantar pressure parameters or contact area during gait.

Ligamentous laxity

The plantar flexion range of motion differences that are seen between men and women may be reflective of ligamentous laxity. Many studies have suggested that women have greater knee and ankle laxity values [14–18]. Beynnon and colleagues [6] hypothesized that serum estradiol and progesterone levels were associated with increased ankle and knee joint laxity. Using validated methods, a KT1000 arthrometer, and TLOS ankle ligament testing device, the ligamentous laxity was tested in 17 women and 17 male controls. This study demonstrated increased ligamentous laxity in women as compared with men in the knee and ankle; however, there was no variation of laxity seen over the menstrual cycle.

Extrinsic factors

Many extrinsic factors have been implicated in sports injury for men and women. These include training intensity, terrain, sport type, competition level, and practice and game surfaces to name a few. The most extensively studied extrinsic factor that causes injury for women is the shoe. Shoe structure and shape has been driven by fashion and not by foot comfort and appropriate fitting. High-heeled, narrow toe-box shoes result in excessively high plantar pressures and toe crowding, which result in hammer toes, neuromas, bunionettes, and hallux valgus. A national health interview survey reported that 1% of all adults have hallux valgus. The prevalence increases as people age, with hallux valgus involving 9% of 30- to 60-year-olds and 16% of people older than 60 years of age. Women are two to four times more likely to develop hallux valgus than are men. Hammer toes are four to five times more common in women and affect 2% to 20% of the population; many of these people will require surgery. It was estimated that 209,000 bunionectomies, 210,000 hammertoe corrections, 66,500 neuroma resections, and 119,000 bunionette repairs were performed in 1991. The economic impact was 1.5 billion dollars for direct costs and 1.5 billion dollars for indirect costs (eg, time off work). Assuming that most patients have forefoot deformity related to shoe wear, the cost to society for 1 year for poorly fitting shoes is approximately 3 billion dollars [19,20].

Increasing heel height is believed to be a major contributor in forefoot pathology. Studies have demonstrated that women who wear shoes with high heels complain of foot pain and deformity more often than do women who wear flats or are barefoot [21,22]. High heels lead to increased peak pressures beneath the metatarsal heads, increased rate of plantar loading, and decreased time to maximal peak pressure (Fig. 1). The increase in pressure and rate of loading are believed to play a significant role in forefoot pathology by increasing the amount of stress to the forefoot. Shoes of increasing heel height shift the plantar pressure from the heel and midfoot to the medial forefoot. A heel height greater than 25 mm has been implicated in hallux valgus and plantar calluses [23,24].

Narrow, pointed shoes also play a significant role in forefoot deformity. Ancient skeletal remains have been evaluated and illustrate that hallux valgus occurred in later medieval burials only after introduction of the narrow toe-boxed shoe [25]. It is believed that as women consistently wear tight shoes with a narrow toe box, they slowly deform the foot to fit the shoe. In a study of women and their shoe wear, 88% of women wore shoes that were smaller in width than their feet [20]. The larger the foot–shoe discrepancy, the greater the incidence of foot pain and deformity. The most common forefoot pathology that was seen in these patients was hallux valgus (in 71% of subjects). Other forefoot complaints, such as corns and calluses, also are seen more often in people who wear narrow, pointed shoes.

Ill-fitting shoes affect more than just the foot. Injury to other joints, such as hips, knees, and ankles, can occur with poorly fitting shoes (Table 2). Painful feet increase the risk for falls and hamper mobility, which lead to many forms of injury (eg, ankle sprains or fractures). The increase in falls with specific types of shoes is especially important in elderly patients who are osteopenic and fracture more readily. Several studies have evaluated shoe wear at the time of fall in the elderly population, and found that most patients were wearing slippers [27]. The lack of fixation, flexible heel counters, and excessively flexibility of the shoes are associated with a higher incidence of tripping and falls. Because hip fracture is a significant cause of morbidity and mortality in the elderly, the type and stability of the shoe needs to be taken into consideration. It also was shown that heel heights of 1.5 inch or greater increase knee torque with walking. This increased torque is believed to be associated with the increase in knee osteoarthritis that is seen in women who wear high heels [26]. Other studies [31] showed that as the heel height increases, so do the body's oxygen and energy consumption.

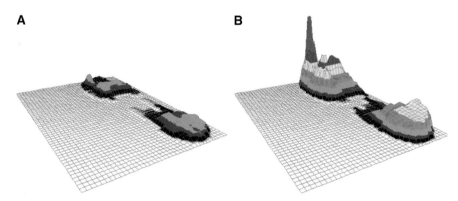

Fig. 1. (*A*) Plantar pressure distribution of the foot in a flat shoe. The plantar surface has little variation in pressures over the forefoot in stance. (*B*) Plantar pressure distribution of the foot in a 2.5-inch heel. The plantar peaks over the forefoot demonstrate the excessive pressures in this region.

Table 2
Influence of shoes on injury

Anatomic location	Reference	Summary
Hip	Sherrington & Menz et al [26]	Older patients who have had a fall-related hip fracture were wearing inappropriate shoes. The most common shoes worn during falls were slippers or shoes without fixation.
Knee	Kerrigan et al [27]	Shoes with moderately high heels (1.5 in) significantly increase knee torque and are believed to be relevant in the formation and progression of knee OA.
Foot	Menz & Morris et al [23]	Narrow shoes are associated with corns, hallux valgus, and foot pain. Shoes shorter than the foot are associated with lesser toe deformity. Heel elevation greater than 25 mm is associated with hallux valgus and plantar calluses.
	Burns et al [28]	Ill-fitting shoes are associated with pain and ulcer formation.
	Snow & Williams et al [24]	Increasing heel height increases the maximum peak pressure under the metatarsal heads in the forefoot, decreases the time to maximum peak pressure under the metatarsal heads, and increases the rate of loading to the metatarsals during early support.
	Frey et al [7]	Women who wear shoes that were too small for their feet have increased pain and deformity.
	Rudicel [22]	Shoes lead to pain and deformity of the foot.
	Hong et al [21]	High heel shoes lead to decreased comfort.
	Corrigan et al [29]	Increasing heel height leads to increased load on the medial forefoot and increased pressure.
Other	Tencer et al [30]	Shoes with a higher heel and less contact area were associated with increased falls in older patients.
	Ebbeling et al [31]	Heart rate and oxygen consumption increase with increased heel height.

Abbreviation: OA, osteoarthritis.

Summary

Limited literature has been published on sexual dimorphism in the foot and ankle. Most articles have focused on injury occurrence and then retrospectively examined differences in the male and female anatomy to explain the injuries that are unique to women. The limited studies that have examined sexual dimorphism and foot and ankle osteology, foot shape, gait, musculature, range of motion, gait parameters, and ligamentous laxity were presented. There are several intrinsic factors that have not been studied in the foot regarding gender difference, including proprioception differences and muscle reaction time. Several investigators have suggested that proprioceptive differences are partially responsible for ankle ligament injuries for men and women. As knowledge of sexual dimorphism and the unique health care needs of women advance, interventions that are related to intrinsic differences, as well as the extrinsic factors of the foot and ankle, will be important to provide appropriate preventative and therapeutic care for women.

References

[1] Wunderlich RE, Cavanagh PR. Gender differences in adult foot shape: implications for shoe design. Med Sci Sports Exerc 2001;34:605–11.

[2] Fessler DMT, Haley KJ, Roshni DL. Sexual dimorphism in foot length proportionate to stature. Ann Hum Biol 2005;32(1):44–59.

[3] Ashizawa K, Kumakura C, Kusumoto A, et al. Relative foot size and shape to general body size in Japanese, Filipinas and Japanese with special reference to habitual footwear types. Ann Hum Bio 1997;24:117–29.

[4] Smith SL. Attribution of foot bones to sex and population groups. J Forensic Sci 1997;42(2):186–95.

[5] Steele DG. The estimation of sex on the basis of the talus and calcaneus. Am J Phys Anthropol 1976;45: 581–8.

[6] Beynnon BD, Bernstein IM, Belisle A, et al. The effect of estradiol and progesterone on knee and ankle joint laxity. Am J Sports Med 2005;33(10):1298–304.

[7] Frey C, Thompson F, Smith J. Update on women's footwear. Foot Ankle Int 1995;16(6):328–31.

[8] Eckstein F, Siedek V, Glaser C, et al. Correlation and sex differences between ankle and knee cartilage morphology determined by quantitative magnetic resonance imaging. Ann Rheum Dis 2004;63: 1490–5.

[9] Sepic SB, Murray MP, Mollinger LA, et al. Strength and range of motion in the ankle in two age groups of men and women. Am J Phys Med 1986;65(2): 75–83.

[10] Endo M, Ashton-Miller JA, Alexander NB. Effects of age and gender on toe flexor muscle strength. J Gerontol 2002;57A(6):M392–7.

[11] Cho SH, Park JM, Kwon OY. Gender differences in three dimensional gait analysis data from 98 healthy Korean adults. Clin Biomech (Bristol, Avon) 2004; 19(2):145–52.

[12] Kerrigan DC, Johansson JL, Bryant MG, et al. Moderate-healed shoes and knee joint torques relevant to the development and progression of knee osteoarthritis. Arch Phys Med Rehabil 2005;86(5): 871–5.

[13] Murphy DF, Beynnon BD, Michelson JD, et al. Efficacy of plantar loading parameters during gait in terms of reliability, variability, effect of gender and relationship between contact area and plantar pressure. Foot Ankle Int 2005;26(2):171–9.

[14] Huston LJ, Wojtys EM. Neuromuscular performance characteristics in elite female athletes. Am J Sports Med 1996;24(4):427–36.

[15] Rosene J, Fogarty T. Anterior tibial translation in collegiate athletes with normal anterior cruciate ligament integrity. J Athl Train 1999;34:93–8.

[16] Rozzi SL, Lephart SM, Gear WS, et al. Knee joint laxity and neuromuscular characteristics of male and female soccer and basketball players. Am J Sports Med 1999;27:312–9.

[17] Beck TJ, Ruff CB, Shaffer RA, et al. Stress fracture in military recruits: gender differences in muscle and bone susceptibility factors. Bone 2000;27(3):437–44.

[18] Wilkerson RD, Mason MA. Differences in men's and women's mean ankle ligamentous laxity. Iowa Orthop J 2000;20:46–8.

[19] Dunn JE, Link CL, Felson DT, et al. Prevalence of foot and ankle conditions in a multiethnic community sample of older adults. Am J Epidemiol 2004; 159(5):491–8.

[20] Frey C. Foot health and shoewear for women. Clin Orthop Relat Res 2000;372:32–44.

[21] Hong WH, Lee YH, Chen HC, et al. Influence of heel height and shoe insert on comfort perception and biomechanical performance of young female adults during walking. Foot Ankle Int 2005;26(12): 1042–8.

[22] Rudicel SA. The shod foot and its implications for American women. J South Orthop Assoc 1994; 3(4):268–72.

[23] Menz HB, Morris ME. Footwear characteristics and foot problems in older people. Gerontology 2005; 51(5):346–51.

[24] Snow RE, Williams KR. High heeled shoes: their effect on center of mass position, posture, three-dimensional kinematics, rearfoot motion, and ground reaction forces. Arch Phys Med Rehabil 1994;75(5):568–76.

[25] Mays SA. Paleopathological study of hallux valgus. Am J Phys Anthropol 2005;126(2):139–49.

[26] Sherrington C, Menz HB. An evaluation of footwear worn at the time of fall-related hip fracture. Age Ageing 2003;32(3):310–4.

[27] Kerrigan DC, Todd MK, Croce UD. Gender differences in joint biomechanics during walking: Normative study in young adults. Amer J Phys Med & Rehab 1998;77(1):2–7.

[28] Burns SL, Leese GP, McMurdo ME. Older people and ill fitting shoes. Postgrad Med J 2002;78(920): 344–6.

[29] Corrigan JP, Moore DP, Stephens MM. Effect of heel height on forefoot loading. Foot Ankle 1993; 14(3):148–52.

[30] Tencer AF, Koepsell TD, Wolf ME, et al. Biomechanical properties of shoes and risk of falls in older adults. J Am Geriatr Soc 2004;52(11):1840–6.

[31] Ebbeling CJ, Hamill J, Crussemeyer JA. Lower extremity mechanics and energy cost of walking in high-heeled shoes. J Orthop Sports Phys Ther 1994;19(4):190–6.

Female Athlete Triad and Stress Fractures
David Feingold, MD[a], Sharon L. Hame, MD[a,b],*

[a]Division of Sports Medicine, Department of Orthopaedic Surgery, The David Geffen University of California Los Angeles School of Medicine, 10833 Le Conte Avenue, Los Angeles, CA 90095, USA
[b]The Greater Los Angeles Veteran's Administration Hospital, 11801 Wishire Boulevard West, Los Angeles, CA 90073, USA

Female athlete triad

The female athlete triad was defined in 1993 by Yeager and colleagues [1]. The three components of the triad are eating disorders, amenorrhea, and osteoporosis [1,2]. Rates of this disorder vary among different groups of female athletes, with the highest rates (up to 62% [3]) in athletes who participate in sports in which appearance is judged (eg, figure skating, gymnastics, ballet) or a low body fat percentage is beneficial (eg, distance running) [4].

Eating disorders

Eating disorders occur in male and female athletes. There is a broad range of disorders that include anorexia nervosa and bulimia nervosa. It has been reported that eating disorders occur more commonly in female athletes, particularly those who participate in aesthetic or weight-dependent sports. Male athletes also are at risk in these types of sports, but they may be more at risk in sports such as wrestling and horse racing [5]. Sundgot-Borgen and Torstveit [6] reported the prevalences of eating disorders to be 22% in men in antigravitation sports, and 42% in female athletes in aesthetic sports. The important component that is involved in eating disorders seems to be insufficient caloric intake for the energy expended, often done as a method of weight loss. This factor affects male and female athletes, but may be more influential in female athletes as demonstrated by the prevalence of the female athlete triad.

Amenorrhea

Amenorrhea is defined as an absence of menses for more than three consecutive cycles. Athletes may have primary or secondary amenorrhea. Amenorrhea in the athletic population may be multifactorial. Any factor that may influence the hypothalamus-pituitary-ovarian axis can affect menses. Such factors include energy deficit, low leptin (a protein hormone that is involved in metabolism and reproductive function) during caloric restriction, and the physical and psychologic stress of training and competing. Energy deficit decreases luteinizing hormone pulse frequency [7]. Low leptin levels may suppress the hypothalamic release of GnRH [8], which controls the secretion of pituitary gonadotropins. The suppression of GnRH markedly diminishes the release of ovarian estrogen. The consequences of untreated hypoestrogenemic amenorrhea include infertility, osteoporosis, cardiovascular disease, and the failure to go through puberty if the disorder arises early in life. It is not surprising that amenorrhea is reported to be more prevalent in the athletic population (3%–66%) than in the general female population (2%–5%) [9]. In men, a decrease in bone mass has been reported in those who have a history of constitutional delay of puberty [10]. The mechanism for this is not understood fully and conflicting data exist [11].

Osteoporosis

Eating disorders, estrogen deficiency, and menstrual dysfunction predispose women to the

* Corresponding author. Department of Orthopaedic Surgery, 10833 Le Conte Avenue, Los Angeles, CA 90095.
 E-mail address: shame@mednet.ucla.edu (S.L. Hame).

third component of the triad: osteoporosis. Osteoporosis is defined as a bone mineral density (BMD) score of less than 2.5 standard deviations below the mean for age. Osteopenia is defined as a BMD score between 1.0 and 2.5 standard deviations below the mean for age. The initial focus of these definitions was to identify patients, primarily in the geriatric age group, who were at increased risk for fracture. Some investigators have changed the focus for the diagnosis of female athlete triad from osteoporosis to osteopenia [12,13]. A study by Pettersson and colleagues [14] in 1999 demonstrated that 10% of female distance runners had osteoporosis, yet nearly 50% were diagnosed with osteopenia. Lower bone density also has been reported in male athletes and it may be a risk factor for stress fracture. The mechanism of decreased bone density in male athletes is not understood well. Research is focused on the role of testosterone and energy deficit on bone in men.

Hormonal influence on bone mineral density

Bone can be influenced dramatically by the actions of hormones. Girls and boys demonstrate androgen receptors in the osteoblasts of their growth plates, which are believed to mediate the anabolic effects of testosterone in bone [15]; however, estrogen was shown to be a more influential hormone with respect to bone. Lower estrogen levels that are seen in the amenorrheic population lead to the loss of the protective effects that this hormone has on bone. Estrogen decreases calcium resorption leading to increased calcium storage within bone. Peak bone mass in females occurs in early adulthood, and the more time spent without menses during growth leads to a lower peak bone mass, ultimately leading to a lower lifetime bone density. Many investigators believe that reduced BMD in premenopausal women seems to be irreversible, despite weight gain, resumption of menses, or estrogen replacement [16,17]. In 1990, Myburgh and colleagues [18] showed a direct correlation between the time spent amenorrheic and the number of stress fractures in athletes. Low estrogen states that are associated with amenorrhea also were shown to have a more profound effect on cancellous bone than on cortical bone [19]. As a result, the incidence of stress fractures in the pelvis, sacrum, and femoral neck are higher in female athletes than male athletes [20,21].

Inadequate calorie intake seems to be the primary mechanism that predisposes female athletes to menstrual dysfunction and resulting detrimental effects on bone [22]. Its role in men is much less understood. Bachrach and colleges [23] reported that nearly 75% of adolescent girls who have anorexia have a BMD that is more than two standard deviations below the normal value. An additional study showed that women who have anorexia nervosa are at an increased risk for stress fracture development [24]. Again, men who have anorexia nervosa and its association with stress fractures have not been studied well.

The overall effect of oral contraceptive pills (OCPs) is not understood well. Although some studies showed that OCPs increased BMD in amenorrheic and anorexic women compared with similar patients who were not on OCPs [25–27], others did not show a change [28,29]. A recent study by Hartard and colleagues [30] showed that OCP use was associated with a decreased BMD of the spine and femoral neck in female endurance athletes. Although OCPs provide estrogen replacement and result in normalization of menses for athletes who are amenorrheic, their overall effects on BMD have yet to be determined.

Stress fractures

More than 95% of all stress fractures occur in the lower extremity, which reflects the high repetitive loads that typically are experienced by a weight-bearing bone. The most common sites are the tibia, metatarsals, and fibula [31,32]. In the female collegiate athlete, the incidence of stress fractures is highest in track and field participants [33]. Male track and field athletes also have a higher incidence of stress fractures, whereas the incidence of stress fractures in other sports varies with sex (Fig. 1) [33]. Stress fractures do occur in the upper extremity, especially in participants in racquet and overhead throwing sports [34,35].

Stress fractures are defined as skeletal defects that result from repeated application of stress that is less than that required to fracture a bone in a single loading, but greater than the bone's ability to recover fully. Microdamage occurs when bone fails to remodel adequately with the application of repetitive, subthreshold stress. Although rare in the general population, the annual

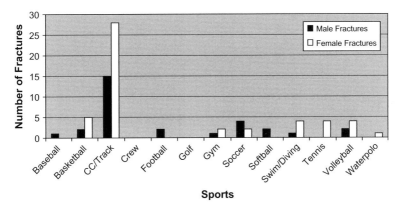

Fig. 1. The total number of primary stress fractures by sport and gender. CC, cross country. (*From* Hame SL, LaFemina J, McAllister DR, et al. Fractures in the collegiate athlete athlete. American J Sports Med 2004;32:449; with permission.)

incidence of stress fractures has been reported to be as high as 20% in young female athletes and military recruits [36], and they account for up to 50% of all injuries that are sustained by runners [37]. In a recent study of 5900 Division 1 collegiate athletes, there was no significant difference in the incidence rate of all fractures between men and women (0.0438 and 0.0461, respectively); however, the incidence of stress fractures was nearly double in women [33]. Distance runners and track athletes seem to be particularly prone to developing stress fractures.

A stress injury to bone may be the result of multiple mechanisms. Extrinsic and intrinsic factors may contribute to injury in male and female athletes. In male athletes and military recruits, the bone commonly experiences excessive strain with accumulation of microdamage and is unable to keep up with repair processes. Other factors also may influence the military population. Valimaki and colleagues [38] reported on the risk factors for stress fractures in military recruits. They found that height, femoral neck and total hip bone mineral content, BMD, and high parathyroid hormone levels were associated with stress fracture incidence. They found no association with sex hormone levels in their population. As for the athletic population, few studies have found significant risk factors for stress fractures in male athletes.

Depressed bony remodeling potential in response to normal strain levels also can occur. Cancellous bone, in which the turnover rate is four times higher than in cortical bone, is affected first. This mechanism plays a major role in patients who have the female athlete triad, metabolic bone disease, or osteoporosis, and it may help to explain the increased occurrence of stress fractures in areas of high cancellous bone content, such as the femoral neck.

In addition to excessive strain and depressed bone remodeling potential, inadequate protective effect of muscle contraction may play a role in stress fracture development. With muscle contraction acting as a shock absorber, muscles can decrease the cortical bending strain that bone experiences. Therefore, muscle fatigue may allow greater strain to be experienced by the cortical surface. Some investigators have stated that differences in neuromuscular control may be a factor in the amount of stress that a bone experiences between men and women [39].

Bone morphology also may play a role in the development of stress fractures. Particularly in the tibia, polar moments of inertia, section modulus, and anterior–posterior and medial lateral widths may have some effect on stress fracture development. A study using male tibias concluded that the characteristics of bone were related to the size of the bone, and that under extreme loading conditions, this might influence the risk for stress fracture [40]. Further study in this area is important and warranted.

In the athlete who has the female athlete triad, stress fractures likely are a result of a combination of mechanisms, but the hormonal factors and osteopenia place these athletes at high risk. It has been hypothesized that estrogen deprivation increases the normal set point for bone to begin remodeling [41,42]. In the hypoestrogenic state, the normal cellular responses to repair damage to the bone are not activated until a greater amount of microdamage has been done. The actual mechanism for this is unclear.

Extrinsic factors, including the type of training program and equipment used, can play important roles in the development of stress fractures. Significant increases in activity often are a triggering event for many stress injuries and explain the high rate of stress fractures among military recruits who are at "boot camp." Equipment, such as footwear, plays a role in force transmission to the weight-bearing bones. It has long been believed that the more rigid the surface on which the athlete trains and competes, the more stress is delivered to a bone. Several studies reported that hard surfaces are not associated with an increased risk for overuse injuries [43–45]. A study by Voloshin [46] reported that running on a grass surface absorbed less shock than did an asphalt surface, therefore potentially being more detrimental than the rigid asphalt surface. Sex differences in the development of stress fractures with respect to the equipment being used and training has yet to be determined.

Diagnosis

Diagnosing stress fractures in men and women requires a high degree of suspicion, a detailed patient history, physical examination, and imaging. Correct diagnosis in the early phase of injury allows for effective treatment and avoids long-term complications. For femoral stress fractures in athletes and military recruits, a surprisingly high percentage (18%–51%) is diagnosed after displacement occurs [47,48]. Athletes usually complain of an insidious onset of activity-related pain at the site of the fracture. A common symptom among most athletes who have stress fractures is worsening pain with activity. Physical examination often can be deceiving. Tenderness and swelling may be noticeable over the fracture. Palpable callus formation in chronic cases also may be seen, particularly in subcutaneous areas. The general presentation of a stress fracture does not seem to be influenced by sex.

All patients who have multiple stress fractures should undergo some routine laboratory testing, including thyroid-stimulating hormone, complete blood count, and routine chemistry panel. These athletes also should have a dual-energy x-ray absorptiometry scan to evaluate BMD. Athletes who are suspected of having female athlete triad should undergo additional laboratory testing, including follicle-stimulating hormone, luteinizing hormone, estradiol, urine pregnancy (if amenorrheic), testosterone, and dexamethasone-suppression testing.

Imaging

Because of the vague nature of the history and physical examination, imaging plays a key role in stress fracture diagnosis for men and women. Often, plain radiographs are normal early in the course of stress fractures. High-resolution radiographs are important to detect the subtle osseous changes that occur as the bone attempts to repair itself. In the acute period, a faint radiolucency in the cortical bone often is the first sign on radiographs. As remodeling occurs, the endosteum becomes thickened and sclerotic. Stress fractures in cancellous bone usually present with a linear band of sclerosis perpendicular to the trabeculae [49,50]. Plain radiographs may take up to 2 to 3 weeks after onset of symptoms to demonstrate callus formation and early fracture healing. In patients who have the female athlete triad, with the inherent disturbance in bone healing, there is likely to be further delay until signs of stress fractures appear on radiographs. Because of this delay, other imaging techniques play an important role. Nuclear medicine scintigraphy (bone scans) is sensitive for diagnosing early stress remodeling and fractures. Acute stress fractures show increased uptake in all three phases of a technetium 99m diphosphonate bone scan [51]. As healing occurs, phase 1 of the scan (flow phase) and then phase 2 (blood pool phase) will normalize. It may take months to years for phase 3 (delayed phase) to normalize as the bone continues to remodel. Although bone scans are sensitive for the evaluation of bone turnover, they are not specific for stress fractures; therefore, they will be positive in the setting of infection, inflammation, tumors, or trauma, all of which may have a similar history and physical examination.

When initial radiographs are negative, MRI often is the next step. MRI is sensitive at detecting endosteal marrow edema, which is one of the earliest signs of stress remodeling [52]. MRI also can show the status of the intramedullary bone, periosteal reaction, and frank fracture lines while providing excellent images of difficult-to-radiograph bones, such as the sacrum (Fig. 2). Arendt and colleagues [53] proposed a grading system that helps to predict time to return to play (Table 1).

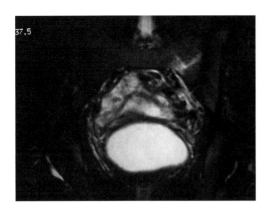

Fig. 2. Sacral stress fracture seen on MRI.

Treatment

When a stress injury is suspected, early treatment steps should be taken. There is no evidence to support sex differences in the specific treatment for stress fractures. Activity modification should be initiated at the first suspicion of the diagnosis. While further imaging studies are being performed, a period of rest should be started to allow time for the initiation of healing processes within the bone. Typically, weight bearing is limited during the initial rest period, but no prospective, randomized studies have revealed how weight bearing affects the rate of healing. Depending on the bone involved in the stress fractures, different cross-training activities, such as cycling or swimming, may allow the athlete to maintain some overall fitness during the early rest period. A supervised, graduated physical therapy program is needed to define the duration, frequency, and intensity of activities as the athlete progresses from rest back to training and eventual return to competition. Location of the fracture and duration of symptoms influence the amount of time needed to heal the fracture. For femoral neck and sacral stress fractures, it takes approximately 6 to 8 weeks of rest for healing to occur [54,55]; however, in patients who have significant nutritional imbalance and osteopenia, this time frame may be longer, and repeated physical examinations and imaging are necessary.

For female athletes who have multiple stress fractures, a high degree of suspicion of the female athlete triad is warranted. For these athletes, treating the causes that led to the triad are of critical importance to allow for healing of the stress fracture and a healthy return to competition. Every female athlete who is suspected of having a stress fracture must undergo a detailed history that focuses on training patterns, eating behavior, and menstrual cycle, which allows for accurate diagnosis of the female athlete triad. Male athletes also must be questioned regarding eating behavior and training patterns.

Delayed diagnosis can lead to eventual displacement of a fracture, and the subsequent poor results following management of these fractures [56,57]. The treatment of stress fractures involves addressing the intrinsic and extrinsic causes of this injury. Extrinsic factors, such as the training program intensity and footwear, should be addressed. In the female athlete triad, those patients who are diagnosed with a stress fracture can be particularly difficult to treat. Diet is an extrinsic factor that plays a significant role in the female athlete triad; it is of critical importance to ensure that the athlete is obtaining enough caloric intake to support her energy output, as well as to help heal the fracture. Stress injuries and fractures require an increase in the normal anabolic rate, which requires increased caloric intake from the normal baseline for an athlete. Athletes—male or female—who have significant eating disorders

Table 1
Grading of stress fracture: MR imaging and plain radiography

Grade	STIR signal change	T2 signal change	T1 signal change	Plain radiograph	Average time to return to play (wk)
1	Present	None	None	Negative	3.3
2	Present	Present	None	Negative	5.5
3	Present	Present	Present	Periosteal reaction	11.4
4	Present	Fracture line	Fracture line	Periosteal reaction or fracture line	14.3

Abbreviation: STIR, short tau inversion recovery.
Adapted from Arendt E, Agel J, Heikes C, et al. Stress injuries to bone in college athletes: a retrospective review of experience at a single institution. Am J Sports Med 2003;31(6):961. *From* Weiss-Kelly A, Hame SL. Stress fractures—returning the athlete to play. Journal of Musculoskeletal Medicine 2005;469; with permission. © 2005 CMP Healthcare Media LLC.

may be especially difficult to treat, and their nutritional imbalance causes an intrinsic inability to heal the microdamage that is occurring. Psychologic assessment may be needed so that dietary habits may be addressed appropriately.

Evaluation of the nutritional status for these athletes is important, including vitamin and mineral intake. The importance of supplemental calcium (1500–2000 mg/d) and vitamin D are well established for patients with nutritional deficiencies, and copper and magnesium play important roles in bone metabolism. Estrogen replacement may help to return the patient to a normal menstrual state and may play a role in improving BMD. For the male athlete with low testosterone, no recommendations regarding the use of testosterone replacement to prevent stress fractures can be made at this time. Bisphosphonates are being evaluated for their role in treating stress fractures. A small study of collegiate level athletes showed a rapid return to training and competition [58]. Another study that investigated a prophylactic role in military recruits did not show a statistical difference between two groups: one used risedronate and the other did not [59]. The current investigations into bisphosphonates may lead to a better understanding of their role in the treatment of these injuries.

Parathyroid hormone (PTH) is another area of research interest for treatment. It plays a significant role in intra- and extracellular homeostasis. In 2002, the US Food and Drug Administration approved teriparatide, a synthetic PTH, for treatment of osteoporosis, based on the results of two randomized studies that showed increased BMD in osteoporotic women who used PTH [60,61]. Although experimental, PTH eventually may play a role in stress fracture treatment by improving healing in slowly resolving stress fractures and allowing an earlier return to competition [62].

Nonsteroidal anti-inflammatory drugs (NSAIDs) are used commonly for treatment of musculoskeletal pain. Theoretically, they may delay fracture healing, which may be detrimental when healing already is compromised in athletes who have the female athlete triad. Based on the current literature, no firm evidence-based recommendation can be made regarding the use of NSAIDs in the treatment of stress fractures [58].

Chronic stress fractures often do not respond to conservative treatment and may require surgical stabilization. Navicular and fifth metatarsal stress fractures have a high incidence of nonunion, and stress fractures that are located on the tension side of a bone (anterior tibia, lateral femoral neck) often fail to heal and are at risk for displacement. Despite the reports of delayed healing in these locations, some studies showed success with conservative treatment. A study of eight basketball players who had anterior tibia stress fracture showed that seven of them healed and returned to full activity at a mean of 12.7 months with conservative treatment that consisted of activity modification and electrical stimulation [63]. Even with this success, there was a significant delay in return to competition, which can be detrimental to the high-level athlete. Surgical intervention, therefore, may lead to an overall improved outcome (Fig. 3).

Femoral neck stress fractures deserve special consideration. As compared with male athletes, female athletes have a higher incidence of these fractures. When located on the tension side of the femoral neck (lateral femoral neck) they have a poor rate of healing with nonsurgical treatment. Typically, these fractures are treated with open reduction and internal fixation.

Complications

Complications are associated with nonoperative and surgical management of stress fractures. Nonoperative management may result in displacement of the fracture, as well as nonunion, malunion, and osteonecrosis [52,64]. Lee and colleagues [65] showed that 23.8% of patients who had a displaced femoral neck fracture

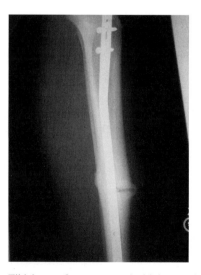

Fig. 3. Tibial stress fracture treated with intermedullary nailing.

developed avascular necrosis of the femoral head. Surgical treatment carries with it all of the well-documented risks for surgery. In addition, patients are still at risk for developing a nonunion, malunion, osteonecrosis, and arthritic changes, as well as displacement during attempted internal fixation.

Return to play

Often, return-to-play decisions are difficult for the physician who treats highly competitive athletes. Multiple investigators have suggested a classification of stress fractures into low- and high-risk categories [66–68]. Low-risk fractures include the femoral shaft, medial tibia, ribs, ulna shaft, and first through fourth metatarsals. High-risk stress fractures include the femoral neck, patella, anterior tibial diaphysis, the talus, the tarsal navicular, proximal fifth metatarsal, and first metatarsal phalangeal sesamoids [69]. An in-depth discussion between the practitioner, athlete, athletic trainer, and coach should be undertaken with emphasis on everyone understanding the potential risks that are involved with undertreatment or overtreatment of the stress fracture. This allows every participant in the athlete's care to understand the importance of a gradual return to activity, with multiple repeat examinations by the physician. Regardless of the category of stress fracture, asymptomatic full weight bearing, no tenderness with palpation over the involved area of injury, and addressing the intrinsic and extrinsic factors of the female athlete triad are required before any return to a high level of training or competition.

Prevention

All high-level athletes should undergo a pre-participation examination by a medical professional who is trained to detect signs of the female athlete triad or its male counterpart. Coaches and athletic trainers should be aware of the female athlete triad and monitor their athletes for signs and symptoms of the disorder. All athletes should be educated on the importance of dietary habits, calcium intake, training and competition programs, as well as hormonal issues and what influences them.

Summary

Stress fractures are a common occurrence in athletes; they account for 50% of all injuries that are sustained by runners, with an incidence in women that is double that in men among Division 1 collegiate athletes [33]. Differences that are influenced by gender seem to have an important role. Newer information is becoming available with respect to risk factors in male athletes and military recruits, including hormonal factors (eg, testosterone levels).

Further research on stress fractures, the risk factors, sex differences, and hormonal influences are necessary to develop potential prevention strategies for stress fractures, the female athlete triad, and its male counterpart.

References

[1] Yeager K, Agostini R, Nativ A, et al. The female athlete triad: disordered eating, amenorrhea, osteoporosis. Med Sci Sports Exerc 1993;25:775–7.

[2] Otis CL, Drinkwater B, Johnson M, et al. American College of Sports Medicine Position Stand: the female athlete triad. Med Sci Sports Exerc 1997;29: i–ix.

[3] Rosen LW, Hough DO. Pathogenic weight-control behaviors of female college gymnasts. Phys Sportsmed 1988;19:141.

[4] Nattiv A, Agostini R, Drinkwater B, et al. The female athlete triad: the inter-relatedness of disordered eating, amenorrhea, and osteoporosis. Clin Sports Med 1994;13(2):405–18.

[5] Baum A. Eating disorders in the male athlete. Sports Med 2006;36(1):1–6.

[6] Sundgot-Borgen J, Torstveit MK. Prevalence of eating disorders in elite athletes is higher than the general population. Clin J Sports Med 2004;14(1):25–32.

[7] Loucks AB. Energy availability, not body fatness, regulated reproductive function in women. Exerc Sport Sci Rev 2003;31:144–8.

[8] Barash IA, Cheung CC, Weigle DS, et al. Leptin is a metabolic signal to the reproductive system. Endocrinology 1996;137:3144–7.

[9] Otis CL. Exercise-associated amenorrhea. Clin Sports Med 1992;11:351–62.

[10] Finkelstein JS, Neer RM, Biller BMK, et al. Osteopenia in men with a history of delayed puberty. N Engl J Med 1992;326:600–4.

[11] Bertelloni S, Baroncelli GI, Ferdeghini M, et al. Normal volumetric bone mineral density and bone turnover in young men with histories of constitutional delay of puberty. J Clin Endocrinol Metab 1998; 83:4280–3.

[12] Khan KM, Liu-Ambrose T, Sran MM, et al. New criteria for female athlete triad syndrome? Br J Sports Med 2002;36:10–3.

[13] Loucks AB. Introduction to menstrual disturbances in athletes. Med Sci Sports Exerc 2003;35(9):1551–2.

[14] Pettersson U, Stalnacke B, Ahlenius G, et al. Low bone mass density at multiple skeletal sites including

[14] ...the appendicular skeleton in amenorrheic runners. Calcif Tissue Int 1999;64:125–77.
[15] Abu EO, Horner A, Kusec V, et al. The localization of androgen receptors in human bone. J Clin Endrocrinol Metab 1997;82:3493–7.
[16] Drinkwater BL, Nilson K, Ott S, et al. Bone mineral density after resumption of menses in amenorrheic athletes. JAMA 1986;256(3):380–2.
[17] Rigotti NA, Neer RM, Skates SJ, et al. The clinical course of osteoporosis in anorexia nervosa: a longitudinal study of cortical bone mass. JAMA 1991;265(9):1133–8.
[18] Myburgh KH, Hutchins J, Fataar AB, et al. Low bone density is an etiologic factor for stress fractures in athletes. Ann Intern Med 1990;113:754–9.
[19] Slemenda CW, Reister TK, Hui SL, et al. Influences on skeletal mineralization in children and adolescents: evidence for varying effects of sexual maturation and physical activity. J Pediatr 1994;125(2):201–7.
[20] Lombardo SJ, Benson DW. Stress fractures of the femur in runners. Am J Sports Med 1982;10(4):219–27.
[21] Johnson AW, Weiss CB, Stento K, et al. Stress fractures of the sacrum; an atypical cause of low back pain in the female athlete. Am J Sports Med 2001;29(4):498–508.
[22] Pepper M, Akuthota V, McCarty EC. The pathophysiology of stress fractures. Clin Sports Med 2006;25:1–16.
[23] Bachrach LK, Guido D, Katzman K, et al. Decreased bone density in adolescent girls with anorexia nervosa. Pediatrics 1990;86(3):440–7.
[24] Nattiv A, Puffer JC, Green GA. Lifestyles and health risks of collegiate athletes; a multicenter study. Clin J Sport Med 1997;7(4):262–72.
[25] Hergenroeder AC. Bone mineralization, hypothalamic amenorrhea, and sex steroid therapy in female adolescents and young adults. J Pediatr 1995;126(5 Pt 1):683–9.
[26] Cumming DC, Wall SR, Galbraith MA, et al. Reproductive hormone responses to resistance exercise. Med Sci Sports Exerc 1987;19(3):234–8.
[27] Seeman E, Szmukler GI, Formica C, et al. Osteoporosis in anorexia nervosa; the influence of peak bone density, bone loss, oral contraceptive use, and exercise. J Bone Miner Res 1992;7(12):1467–74.
[28] Warren MP, Brooks-Gunn J, Hamilton LH, et al. Scoliosis and fractures in young ballet dancers: relation to delayed menarche and secondary amenorrhea. N Engl J Med 1986;314(21):1348–53.
[29] Kilbanski A, Biller BM, Schoenfeld DA, et al. The effects of estrogen administration on trabecular bone loss in young women with anorexia nervosa. J Clin Endocrinol Metab 1995;80(3):898–904.
[30] Hartard M, Kleinmond C, Kirchbichler A, et al. Age at first oral contraceptive use as a major determinant of vertebral bone mass in female endurance athletes. Bone 2004;35(4):836–41.
[31] Bennell KL, Malcolm SA, Thomas SA, et al. The incidence and distribution of stress fractures in competitive track and field athletes. Am J Sports Med 1996;24:211–7.
[32] Bruckner PD, Bradshaw C, Khan KM, et al. Stress fractures: a review of 180 cases. Clin Sports Med 1996;6:85–9.
[33] Hame SL, LaFemina JM, McAllister DR, et al. Fractures in the collegiate athlete. Am J Sports Med 2004;32(2):446–52.
[34] Jones GL. Upper extremity stress fractures. Clin Sports Med 2006;25:159–74.
[35] Sinha AK, Keading CC, Wadley GM. Upper extremity stress fractures in athletes: clinical features of 44 cases. Clin J Sports Med 1999;9:199–202.
[36] Brukner P, Bennell K, Matheson G. Stress fractures. Victoria (Australia): Blackwell Science; 1999.
[37] McBryde AM. Stress fractures in runners. Clin Sports Med 1985;4:737–52.
[38] Valimaki VV, Alfthan H, Lehmuskallio E, et al. Risk factors for clinical stress fractures in male military recruits: a prospective cohort study. Bone 2005;37(2):267–73.
[39] Hakkinen K. Force production characteristics of leg extensor, trunk flexor and extensor muscles in male and female basketball players. J Sports Med Phys Fitness 1991;31:325–31.
[40] Tommassini SM, Nasser P, Schaffler MB, et al. Relationship between bone morphology and bone quality in male tibias: implications for stress fracture risk. J Bone Miner Res 2005;20(8):1372–80.
[41] Nattiv A, Armsey TD. Stress injury to bone in the female athlete. Clin Sports Med 1997;16(2):197–224.
[42] Frost HM. A new direction for osteoporosis research: a review and proposal. Bone 1991;12(6):429–37.
[43] Hoeberigs JH. Factors related to the incidence of running injuries: a review. Sports Med 1992;13:408–22.
[44] Macaera CA. Lower extremity injuries in runners: advances in prediction. Sports Med 1992;13:50–7.
[45] Marti B. Health effects of recreational running in women: some epidemiological and preventive aspects. Sports Med 1991;11:20–51.
[46] Voloshin KW. Dynamic loading during running on various surfaces. Hum Mov Sci 1992;11:675–89.
[47] Hershman E, Lombardo J, Bergfeld J. Femoral shaft stress fractures in athletes. Am J Sports Med 1990;9:111–9.
[48] Volpin R. Stress fractures of the femoral neck following strenuous activity. J Orthop Trauma 1990;4:394–8.
[49] Daffner RH, Pavlov H. Stress fractures: current concepts. AJR Am J Roentgenol 1992;159:245–52.
[50] Sofka CM. Imaging of stress fractures. Clin Sports Med 2006;25:53–62.
[51] Prather JL, Nusynowitz ML, Snowdy HA, et al. Scintigraphic findings in stress fractures. J Bone Joint Surg Am 1997;59:869–74.

[52] Kiuru MJ, Niva M, Reponen A, et al. Bone stress injuries in asymptomatic elite recruits: a clinical and magnetic resonance imaging study. Am J Sports Med 2005;33(2):272–6.

[53] Arendt E, Agel J, Heikes C, et al. Stress injuries to bone in college athletes: a retrospective review of experience at a single institution. Am J Sports Med 2003;31(6):959–68.

[54] DeFranco MJ, Recht M, Schils J, et al. Stress fractures of the femur in athletes. Clin Sports Med 2006;25:89–103.

[55] Fredericson M, Salamancha L, Bealieu C. Sacral stress fractures, tracking nonspecific pain in distance runners. Phys Sportsmed 2003;31(2):31–42.

[56] Coady C, Micheli L. Stress fractures in the pediatric athlete. Clin Sports Med 1997;16:225–36.

[57] Ernst J. Stress fractures of the femoral neck. J Trauma 1964;4:71–83.

[58] Stewart G, Brunet ME, Manning MR, et al. Treatment of stress fractures in athletes with intravenous pamidronate. Clin J Sport Med 2005;15(2):92–4.

[59] Milgrom C, Finestone A, Novack V, et al. The effect of prophylactic treatment with risedronate on stress fracture incidence among infantry recruits. Bone 2004;35(2):418–24.

[60] Neer RM, Arnaud CD, Zanchetta JR, et al. Effect of parathyroid hormone (1–34) on fractures and bone mineral density in postmenopausal women with osteoporosis. N Engl J Med 2001;334:1434–41.

[61] Body JJ, Gaich GA, Scheele WH, et al. A randomized double-blind trial to compare the efficacy of teriparatide [recombinant human parathyroid hormone (1–34)] with alendronate in postmenopausal women with osteoporosis. J Clin Endocrinol Metab 2002;87:4528–35.

[62] Koester MC, Spindler KP. Pharmacologic agents in fracture healing. Clin Sports Med 2006;25:63–73.

[63] Rettig A, Shelbourne KD, McCaroll JR, et al. The natural history and treatment of delayed union stress fractures of the anterior cortex of the tibia. Am J Sports Med 1988;16:250–5.

[64] Visuri T. Stress osteopathy of the femoral head. Acta Orthop Scand 1997;68:138–41.

[65] Lee C, Huang G, Chao K, et al. Surgical treatment of displaced stress fractures of the femoral neck in military recruits: a report of 42 cases. Arch Orthop Trauma Surg 2003;123:527–33.

[66] Boden BP, Osbahr DC, Jimenez C. Low-risk stress fractures. Am J Sports Med 2001;29:100–11.

[67] Boden BP, Osbahr DC. High risk-stress fractures: evaluation and treatment. Am Acad Orthop Surg 2000;8(6):344–53.

[68] Brukner P, Bradshaw C, Bennell K. Managing common stress fractures: let risk level guide treatment. Physician Sports Med 1998;26(8):39–47.

[69] Diehl JJ, Best TM, Kaeding CC. Classification and return-to-play considerations for stress fractures. Clin Sports Med 2006;25:17–28.

Anterior Cruciate Ligament Biology and Its Relationship to Injury Forces

James R. Slauterbeck, MD[a], John R. Hickox, MS[b,c], Bruce Beynnon, PhD[a], Daniel M. Hardy, PhD[b,c],*

[a]*Department of Orthopaedics and Rehabilitation, University of Vermont College of Medicine, 95 Carrigan Drive, Stafford Hall, Burlington, VT 05405, USA*
[b]*Department of Cell Biology and Biochemistry, Texas Tech University Health Sciences Center, 3601 Fourth Street, Lubbock, TX 79430-6540, USA*
[c]*Department of Orthopaedic Surgery and Rehabilitation, Texas Tech University Health Sciences Center School of Medicine, 3601 Fourth Street, Lubbock, TX 79430-6540, USA*

The anterior cruciate ligament (ACL) is one of primary stabilizers of the knee, and disruption of this ligament is three to ten times more common in female athletes than male athletes who participate in the same sports at the same level of competition [1–5]. These knee injuries are a growing cause for concern because of the association with early-onset, posttraumatic osteoarthritis after severe ligament tears. Despite an abundance of theories about why female athletes are more susceptible to this injury and proposed neuromuscular training programs designed to protect the knee, no definitive causal link has been found among mechanisms of injury, sex, and ACL rupture.

Determining the causes of female susceptibility to ACL rupture has proved difficult for many reasons. First, knee function is complex and unique. The knee is composed of patellofemoral and tibiofemoral joints, and in humans, the only bipedal mammal, the biomechanics of these joints are not only complex but also unique because they have adapted to our upright bipedal gait. Consequently, studies of nonhuman models, although educational and helpful, are subject to criticism until their relevance to the human knee is established. Second, the sex disparity in ACL injury is an anatomic question and a reproductive biology question because of the complex nature of injury, repair, and remodeling. The reproductive processes vary dramatically among the many animal species. Humans' unique reproductive endocrinology further complicates the interpretation of results from studies performed on nonhuman models. Finally, although many explanations have been proposed for the increased incidence of ACL injury in female athletes, few have been tested adequately because many are interrelated and not amenable to isolation and study.

Injury causation was described by Meeuwisse [6] and modified by Bahr and Krosshaug [2] to define ACL injury mechanisms (Fig. 1). In their approaches, injury is an end result of a combination of internal and external risk factors that make predisposed athletes susceptible to injury. The susceptible athlete sustains an injury after an inciting event places the athletes into a situation in which the ACL fails. At the moment the ACL fails, the applied load exceeds the load that the ACL can withstand. The loads placed on the ligament and the loads at which the ligament will fail are central to the approach proposed by Slauterbeck and colleagues [7] and are the focus of this article.

Funding support for this article was provided by Orthopaedic Research and Education Foundation and NIH grant AR-049767.

* Corresponding author. Department of Cell Biology and Biochemistry, Texas Tech University Health Sciences Center, 3601 Fourth Street, Lubbock, TX 79430-6540, USA.
  E-mail address: Daniel.Hardy@ttuhsc.edu (D.M. Hardy).

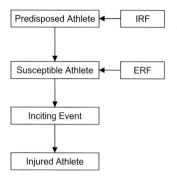

Fig. 1. The basic description of injury models by Bahr [2] and Meeuwisse [6]. IRF, internal risk factors; ERF, external risk factors.

Relationships among anterior cruciate ligament injury factors

Proposed causes of sex differences in the incidence rate of ACL injuries range from extrinsic factors, such as footwear and training, to intrinsic factors, such as anatomy [8–13], neuromuscular control of the leg [14], ligament biomechanics [15,16], ligament laxity [13], and hormonal effects [17–19]. These seemingly distinct, competing factors are interdependent. For example, anatomy influences choice of footwear, and coaching can influence neuromuscular control of the leg. Likewise, sex hormones have an effect on anatomic, neuromuscular, and other intrinsic differences between the sexes.

What are the relationships between and among these interdependent factors, and how do their collective influences translate into a situation whereby ACL injury results? To address this question, we constructed a model (Fig. 2) that places the various factors in a hierarchy and identifies pathways by which they contribute to ACL injury. According to this model, all factors that influence susceptibility to ACL injury ultimately must affect either the mechanical load placed on the ligament or the magnitude of the ligament's inherent "load at failure." When the ligament's ultimate failure load is less than the applied load, ACL injury occurs (Fig. 2).

In this view of ACL injury, the intersegmental load produced across the tibiofemoral joint and the corresponding distribution resisted by the ACL is determined in part by the applied external load on the knee and the knee stiffness. The applied loads to the knee (eg, those produced by the ground reaction force, body weight, anatomic structure, and magnitude and sequence of muscle contraction) combine to create the intersegmental load, which is transmitted across the knee and resisted by the ligaments and articular contact. A longer tibia (eg, an increased moment arm) produces greater intersegmental moments and loads across the knee for a given force applied to the foot. An individual who lands with increased valgus angulation about the knee also experiences an increased valgus moment, which must be resisted by articular contact, the ACL, and other knee ligaments. The application of an internal knee torque or valgus moment combined with anterior shear forces increases the load on the ACL [20]. Athletes who exhibit poor neuromuscular control (eg, medial knee collapse) patterns also may increase the applied load to the knee because of the

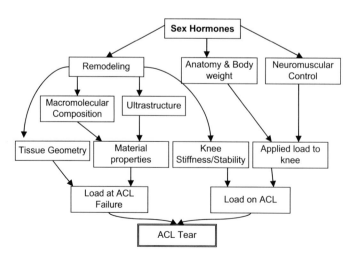

Fig. 2. Relationship of potential factors that affect rate of ACL injury.

position (valgus) or torque (internal) placed on the leg during a landing event [21,22]. The magnitude of the loads ultimately transmitted to the ACL depends in part on the loads applied to the knee (the ground reaction force and a subject's body mass index), the moment arms they act over (eg, an individual's specific anatomy, such as the length of the femur or tibia and lower extremity alignment), and neuromuscular control (eg, the sequence and magnitude of muscle contraction).

Knee stiffness also affects how a load is transmitted to the ACL. If similar loads are applied to a knee with less stiffness, then a greater proportion of the intersegmental loads may be transmitted to the ACL. In humans, knee stiffness in the presence of muscular co-activation is lower in women than men [23]. It is possible that for a given intersegmental load produced across the knee, a greater proportion will be applied directly to the ACL in women compared to men.

The second major determinant of ACL injury is the ligament's intrinsic load at failure. The ACL's geometry (size or shape), macromolecular composition, and ultrastructure (internal organization) all determine the magnitude of the load the ligament can withstand before it fails. In humans, the female ACL is smaller than the male ACL [16]. With respect to failure load, an ACL of smaller cross-sectional area and similar composition fails at lower loads compared with an ACL of larger area. Conversely, an ACL of inferior internal composition or structure (material properties) compared with that is the same size but has superior internal composition or structure fails at lower stresses [16]. The size, shape, and material properties of the ACL depend on how effectively the ligament remodels in response to loads placed on it, and they are important determinants of susceptibility to injury.

Effects of tissue remodeling

Tissue remodeling is an ongoing process whereby old or damaged structures are degraded and replaced with newly synthesized molecules [24–28]. This process has the potential to have an effect on several aspects of ACL injury susceptibility (Fig. 2). First, tissue remodeling determines the size, shape, and internal composition of the ACL. Any difference or change in ACL structure or geometric quality would alter the ligament's intrinsic susceptibility to failure. Second, tissue remodeling can change knee stiffness, defined as resistance to anterior translation of the tibia relative to the femur, by modifying the soft tissues about the knee. Third, a major remodeling event in humans occurs during puberty, when sex hormones directly change the size, shape, and structure of the human body. Differences in primary knee ligaments' (medial collateral ligament [MCL], lateral collateral ligament [LCL], posterior cruciate ligament [PCL]) laxity, knee laxity, femur and tibia length, and neuromuscular control during puberty collectively affect how loads are transmitted to the ACL. Remodeling alters the load at which an ACL fails and the magnitude of the loads applied to the knee and, subsequently, the ACL.

Tissue remodeling occurs continuously in injured and normal tissues. The balance between the degradative and biosynthetic arms of this process is determined by the relative activities of matrix metalloproteinases (MMPs) and tissue inhibitors of metalloproteinases (TIMPs) and by the rate at which new structural components are synthesized [24,25,27,28]. MMPs are endoproteinases that are responsible for matrix breakdown [28], and TIMP proteins block the hydrolytic activities of MMPs [29,30]. Together, MMPs and TIMPs function as molecules that dictate the rate of matrix breakdown during tissue remodeling. In general, the activity of MMPs is greater than TIMPs during the breakdown component of tissue remodeling [26,31–33]. Conversely, repair is favored when TIMP activity increases relative to MMPs. Whether a noninjured tissue gets larger or smaller or stays the same size is determined in part by expression of genes encoding structural proteins, enzymes that degrade them, and inhibitors of those enzymes. Likewise, whether an injured tissue is adequately repaired is determined in part by expression of these genes.

Recent studies have characterized expression of genes that encode tissue remodeling components in the human ACL. MMPs are expressed by many cell types, including macrophages, neutrophils, fibroblasts, trophoblasts, endometrial cells, epithelial cells, and various tumor cells. These enzymes are also present in connective tissues (eg, cartilage) and in synovial fluid [34–36]. Nine MMPs and all four known TIMPs are expressed in uninjured, human ACL tissue [37]. After 6 days in tissue culture, the ACL, medial collateral ligament, and patellar tendon from rabbits also express collagenase (a tissue breakdown enzyme) activity [38]. Immediately after injury, ACLs exhibit increased protease (enzymes that breakdown proteins) activity that decreases during the recovery period [15,34,39]. Some of the increased MMP activity present in the injured ACL is derived from inflammation, but some—if not most—may originate in the ligament itself [39]. These studies collectively demonstrate that MMP and TIMP genes are expressed in healthy ACLs, and their expression changes dynamically after injury.

Tissue remodeling is also evident in the injured ACL and in the tissues used for ACL reconstruction. For example, in sheep ACL subjected to thermal injury, the ligament's size increases and material properties change as the injured tissue is repopulated with new cells [40]. After a tendon is used to reconstruct the injured ACL, the tendon undergoes a remodeling process called "ligamentization" as it transforms from a tendon into a tissue resembling a ligament [39]. Despite the essential roles of normal remodeling events in maintenance and repair of all tissues, including the ACL, their potential contribution to the sex disparity in ACL injury rate has not been evaluated adequately.

Sex, hormones, and anterior cruciate ligament injury

Sex and sex hormones influence several aspects of human ACL function and injury, including effects on the load on the ligament and its intrinsic load at failure (Fig. 2). No sex difference in the incidence rate of ACL injury before puberty has been reported. After puberty, however, the incidence rates of ACL injuries are greater for women compared with men who take part in similar activities or common sports [3,41]. Concomitantly, neuromuscular strategies that control jumping and landing diverge during puberty. After puberty, men continue to land with their knees wide apart and women land with their knees close together, which creates an increased valgus moment about the knee [42]. This factor is a concern because an increased valgus moment has been shown to be a risk factor for increased risk of ACL injury [43]. Anatomically, sex hormones are responsible for the musculoskeletal changes that occur during puberty, including changes in musculoskeletal strength, fat distribution, growth, and shape. Some of these anatomic differences between men and women are considered risk factors for ACL injury [13,44]. Sex-dependent changes in human anatomy and neuromuscular control can influence ACL injury rate through their effect on the load applied to the ligament (Fig. 2).

Several studies have provided evidence that the phase of the menstrual cycle and the associated variations in hormones may be risk factors that influence knee ligament injury among female athletes [4,18,19,45]. Prior work by our group has revealed that the likelihood of sustaining an ACL injury does not remain constant during the menstrual cycle; instead, the risk of suffering an ACL disruption is significantly greater during the preovulatory phase of the menstrual cycle compared with the postovulatory phase (odds ratio, 3.2) [46]. Further evidence in support of the observation that sex hormones affect the likelihood of suffering an ACL injury comes from our research with the canine model, which demonstrated that ACL tears are more prevalent in spayed female and neutered male canines than in their gonadally intact counterparts [47]. Estrogen decreases the failure load of the rabbit ACL [48] but not the sheep ACL [49,50], a discrepancy that may reflect a species difference in the response to sex hormones. Although these studies do not establish a definitive link between sex hormones and ACL injury, they support the hypothesis that such a link exists.

Sex hormones exert their biologic effects almost entirely by affecting the regulation of gene expression. Estrogen and progesterone regulate the transcription of many TIMP and MMP genes [35,51–53]. For example, expression of specific MMPs in cycling human endometrium depends on the phase of the menstrual cycle [35]. Several endometrial MMPs degrade Type I collagen, which is the major structural collagen in the ACL [51,54–56]. Estrogen-dependent degradation of Type I collagen results in dilation of the cervix in the guinea pig during birth [51]. Estrogen-dependent collagenase production and progesterone-dependent inhibition of collagenase have been observed in the pig's pubic ligament [53]. Receptors for testosterone, estrogen, and progesterone are present in the human ACL [57–59]. Increasing the concentration of estrogen in an ACL tissue culture model has been shown to result in a decrease in procollagen production [60]. Collectively, these studies suggest that variation in steroid hormone levels during the menstrual cycle could affect expression of tissue remodeling genes in the ACL, including those for major structural proteins, MMPs, or TIMPs, which could, in turn, affect the intrinsic load at which a ligament fails.

Tissue remodeling genes are differentially expressed between men and women. The average level and range of matrix metalloproteinase-3 (proteases responsible for matrix breakdown) messenger ribonucleic acid (mRNA) expression in the human ACL were found to be higher in women than men [61]. The measured amount of metalloproteinase-3 protein in human ACL correlated highly with expression of its mRNA [61]. Similarly, the ratios of metalloproteinase-3 and matrix metalloproteinases-1 to collagen $1\alpha 1$ gene expression were found to be higher in ACLs obtained from women than from men in ACLs from patients undergoing total knee arthroplasty [62]. These ratios can be viewed as measures of the relative rates of collagen degradation and synthesis. At least three genes responsible for ligament remodeling are differentially expressed between the sexes, and their expression seems to favor ligament degradation in women compared to men.

Summary

The model presented in Fig. 2 illustrates that ACL injury is determined by two variables: the ultimate failure load of the ligament and the mechanical load applied to the ligament. All factors that contribute to ACL injury must do so by affecting one or both of these two basic variables. Some factors, such as sex hormones and tissue remodeling, have a multifaceted effect on the failure load of the ACL and the magnitude of the load applied to this ligament. The relationships defined by this model provide a useful framework for interpretation of results on risk factors for ACL injury and may generate some new questions to be answered by future research in this area. Finally, the model also illustrates the potentially profound effects that sex hormones and tissue remodeling likely have on female susceptibility to ACL injuries.

References

[1] Arendt E, Dick R. Knee injury patterns among men and women in collegiate basketball and soccer: NCAA data and review of literature. Am J Sports Med 1995;23:694–701.

[2] Bahr R, Krosshaug T. Understanding injury mechanisms: a key component of preventing injuries in sport. Br J Sports Med 2005;39:324–9.

[3] Hewitt TE, Lindenfeld TN, Riccobene JV, et al. The effect of neuromuscular training on the incidence of knee injury in female athletes: a prospective study. Am J Sports Med 1999;27:699–706.

[4] Myklebust G, Maehlum S, Holm I, et al. A prospective cohort study of anterior cruciate ligament injuries in elite Norwegian team handball. Scand J Med Sci Sports 1998;8:149–53.

[5] Gray J, Taunton JE, McKenzie DC, et al. A survey of injuries to the anterior cruciate ligament of the knee in female basketball players. Int J Sports Med 1985;6:314–6.

[6] Meeuwisse WH. Assessing causation in sport injury: a multifactorial model. Clin J Sport Med 1994;4:166–70.

[7] Slauterbeck JR, Hickox JR, Hardy DM. Ligament biology and its relationship to injury forces. In: Hewitt T, editor. Understanding and preventing non-contact ACL injury. Rosemont (IL): American Orthopaedic Society for Sports Medicine (AOSSM); 2007.

[8] LaPrade RF, Burnett QM. Femoral intercondylar notch stenosis and correction to anterior cruciate ligament injuries: a prospective study. Am J Sports Med 1994;22:198–203.

[9] Lund-Hanssen H, Gannon J, Engebretsen L, et al. Intercondylar notch width and the risk for anterior cruciate ligament rupture: a case-control study in 46 female handball players. Acta Orthop Scand 1994;65:529–32.

[10] Souryal TO, Moore HA, Evans JP. Bilaterality in anterior cruciate ligament injuries: associated intercondylar notch stenosis. Am J Sports Med 1988;16:449–54.

[11] Souryal TO, Freeman TR. Intercondylar notch size and anterior cruciate ligament injuries in athletes: a prospective study. Am J Sports Med 1993;21:535–9.

[12] Tillman MD, Smith KR, Bauer JA, et al. Differences in three intercondylar notch geometry indices between males and females: a cadaver study. Knee 2002;9:41–6.

[13] Uhorchak JM, Scoville CR, Williams GN, et al. Risk factors associated with non-contact injury of the anterior cruciate ligament: a prospective four-year evaluation of 859 West Point cadets. Am J Sports Med 2003;9:41–6.

[14] Ford KR, Myer GD, Toms HE, et al. Gender differences in the kinematics of unanticipated cutting in young athletes. Med Sci Sports Exerc 2005;37:124–9.

[15] Chandrashekar N, Mansouri H, Slauterbeck J, et al. Sex-based differences in the tensile properties of the human anterior cruciate ligament. J Biomech 2006; Jan 4 (Epub ahead of print).

[16] Chandrashekar N, Slauterbeck J, Hashemi J. Sex-based differences in the anthropomorphic characteristics of the anterior cruciate ligament and its relation to intercondylar notch geometry. Am J Sports Med 2005;33:1492–8.

[17] Slauterbeck JR, Hardy DM. Sex hormones and knee ligament injuries in female athletes. Am J Med Sci 2001;322:196–9.

[18] Slauterbeck JR, Fuzie SF, Smith MP, et al. The menstrual cycle, sex hormones, and anterior cruciate ligament injury. J Athl Train 2002;37:275–8.

[19] Wojtys EM, Huston LJ, Boynton MD, et al. The effect of the menstrual cycle on anterior cruciate ligament injuries in women as determined by hormone levels. Am J Sports Med 2002;30:182–8.

[20] Markolf KL, Burchfield DM, Shapiro MM, et al. Combined knee loading states that generate high anterior cruciate ligament forces. J Orthop Res 1995; 13:930–5.

[21] Hewitt TE, Stroupe AL, Nance TA, et al. Plyometric training in female athletes: decreased impact forces and increased hamstring torques. Am J Sports Med 1996;24:765–73.

[22] Markolf KL, Slauterbeck JR, Armstrong KL, et al. A biomechanical study of replacement of the posterior cruciate ligament with a graft. Part II: Forces in the graft compared with forces in the intact ligament. J Bone Joint Surg Am 1997;79A:381–6.

[23] Wojtys EM, Ashton-Miller JA, Huston LJ. A gender-related difference in the contribution of the knee musculature to sagittal-plane shear stiffness in subjects with similar knee laxity. J Bone Joint Surg Am 2002;84A:10–6.

[24] Dahlberg L. A longitudinal study of cartilage matrix metabolism in patients with cruciate ligament rupture-synovial fluid concentrations of aggrecan fragments, stromelysin-1 and tissue inhibitor of metalloproteinase-1. Br J Rheumatol 1994;33: 1107–11.

[25] DiGirolamo N. Increased matrix metalloproteinases in the aqueous humor of patients and experimental animals with uveitis. Curr Eye Res 1996;15: 1060–8.

[26] Edwards DR. Differential effects of transforming growth factor beta-1 on the expression of matrix metalloproteinases and tissue inhibitors of metalloproteinases in young and old human fibroblast. Exp Gerontol 1996;31:207–23.

[27] Everts V. Phagocytosis and intracellular digestion of collagen, its role in turnover and remodeling. Histochem J 1996;28:229–45.

[28] Gaire M. Structure and expression of the human gene for the matrix metalloproteinase matrilysin. J Biol Chem 1994;269:2032–40.

[29] Edwards DR. The roles of tissue inhibitors of metalloproteinases in tissue remodeling and cell growth. Int J Obes Relat Metab Disord 1996;20(Suppl 3): 9–15.

[30] Mankin HJ, Mow VC, Buckwalter JA. Form and function of articular cartilage. In: Simon SR, editor. Orthopaedic basic science. Rosemont (IL): American Academy of Orthopaedic Surgeons; 1994. p. 1–144.

[31] Kozaci LD. Degradation of type II collagen, but not proteoglycan, correlates with matrix metalloproteinase activity in cartilage explant cultures. Arthritis Rheum 1997;40:164–74.

[32] Pinals RS. Mechanisms of joint destruction, pain and disability in osteoarthritis. Drugs 1996;52 (Suppl 3):14–20.

[33] Salamonsen LA, Woolley DE. Matrix metalloproteinases in normal menstruation. Hum Reprod 1996;11(Suppl 2):124–33.

[34] Lohmander LS. Temporal patterns of stromelysis-1, tissue inhibitor, and proteoglycan fragments in human knee joint fluid after injury to the cruciate ligament or meniscus. J Orthop Res 1994;12:12–8.

[35] Matrisian LM. Matrix metalloproteinase gene expression. Ann N Y Acad Sci 1994;732:42–50.

[36] Meikle MC. Immunolocalization of matrix metalloproteinases and TIMP-1 (tissue inhibitor of metalloproteinase) in human gingival tissues from periodontitis patients. J Periodontal Res 1994;29: 118–26.

[37] Foos MJ, Hickox JR, Mansour PG, et al. Expression of matrix metalloprotease and tissue inhibitor of metalloprotease genes in the human anterior cruciate ligament. J Orthop Res 2001;19:642–9.

[38] Harper J, Amiel D. Collagenase production by rabbit ligaments and tendon. Connect Tissue Res 1988; 17:253–9.

[39] Amiel D. Injury of the ACL: the role of collagenase in ligament degeneration. J Orthop Res 1989;7:486–93.

[40] Jackson DW, Grood ES, Cohn BT, et al. The effect of in situ freezing on the anterior cruciate ligament: an experimental study in goats. J Bone Joint Surg Am 1991;73A:201–13.

[41] Arendt EA. Orthopaedic issues for active and athletic women. Clin Sports Med 1994;13:483–503.

[42] Hewitt TE, Myer GD, Ford KR. Decrease in neuromuscular control about the knee with maturation in female athletes. J Bone Joint Surg Am 2004;86A: 1601–8.

[43] Hewitt TE, Myer GD, Ford KR, et al. Biomechanical measures of neuromuscular control and valgus loading of the knee predict anterior cruciate ligament injury risk in female athletes: a prospective study. Am J Sports Med 2005;33:492–501.

[44] Ireland ML. The female ACL: why is it more prone to injury? Orthop Clin North Am 2002;33:637–51.

[45] Arendt EA, Bershadsky B, Agel J. Periodicity of noncontact anterior cruciate ligament injuries during the menstrual cycle. J Gend Specif Med 2002;5: 19–26.

[46] Beynnon BD, Johnson RJ, Braun S, et al. The relationship between menstrual cycle phase and anterior cruciate ligament injury: a case-control study of recreational alpine skiers. Am J Sports Med 2006;34: 757–64.

[47] Slauterbeck JR, Pankratz K, Xu KT, et al. Canine ovariohysterectomy and orchiectomy increases the prevalence of ACL injury. Clin Orthop Relat Res 2004;429:301–5.

[48] Slauterbeck J, Clevenger C, Lundberg W, et al. Estrogen level alters the failure of the rabbit anterior cruciate ligament. J Orthop Res 1999;17:405–8.

[49] Seneviratne A, Attia E, Williams RJ, et al. The effect of estrogen on ovine anterior cruciate ligament fibroblasts: cell proliferation and collagen synthesis. Am J Sports Med 2004;32:1613–8.

[50] Strickland SM, Belknap TW, Turner SA, et al. Lack of hormonal influences on mechanical properties of sheep knee ligament. Am J Sports Med 2003;31: 210–5.

[51] Rajabi MR. Immunochemical and immunohistochemical evidence of estrogen-mediated collagenolysis as a mechanism of cervical dilatation in the guinea pig at parturition. Endocrinology 1991;128: 371–8.

[52] Schneikert J. Androgen receptor-ets protein interaction is a novel mechanism for steroid hormone-mediated down-modulation of matrix metalloproteinase expression. J Biol Chem 1996;271: 23907–13.

[53] Wahl LW. Effect of hormones on collagen metabolism and collagenase activity in the pubic symphysis ligament of the guinea pig. Endocrinology 1977;100: 571–9.

[54] Aimes RT. Matrix metalloproteinase-2 is an interstitial collagenase. J Biol Chem 1995;270: 5872–6.

[55] Krane SM. Different collagenase gene products have different roles in degradation of type I collagen. J Biol Chem 1996;271:28509–15.

[56] Woo SL, An KN, Arnoczky SP. Anatomy, biology and biomechanics of tendon, ligament and meniscus. In: Simon SR, editor. Orthopaedic basic science. Rosemont (IL): American Academy of Orthopaedic Surgeons; 1994. p. 45–88.

[57] Hamlet WP, Liu SH, Panossian V. Primary immunolocalization of androgen target cells in the human anterior cruciate ligament. J Orthop Res 1998;15: 657–63.

[58] Liu SH. Primary immunolocalization of estrogen and progesterone target cells in the human anterior cruciate ligament. J Orthop Res 1996;14:526–33.

[59] Sciore P, Frank CG, Hart DA. Identification of sex hormone receptors in human and rabbit ligaments of the knee by reverse transcription-polymerase chain reaction: evidence that receptors are present in tissue from both male and female subjects. J Orthop Res 1998;16:604–10.

[60] Liu SH. Estrogen affects the cellular metabolism of the ACL: a potential explanation for female athletic injury. Am J Sports Med 1997;25:704–9.

[61] Hardy DM, Hickox JR, Shepherd SM. Gender differences in expression of MMP3 in human ACL. J Ortho Res Transactions 2002;27:926.

[62] Slauterbeck JR, Hickox JR, Xu KT, et al. Expression of remodeling genes in human anterior cruciate ligament varies by gender. J Ortho Res Transactions 2004;29:1304.

Dimorphism and Patellofemoral Disorders
Elizabeth A. Arendt, MD

Department of Orthopaedic Surgery, University of Minnesota, 2450 Riverside Avenue, Suite R200, Minneapolis, MN 55454, USA

Sex is defined as the classification of living things according to their chromosomal compliment. Gender is defined as a person's self-representation as male or female or how social institutions respond to that person on the basis of his or her gender presentation. One frequently divides the topic of dimorphism into the biologic response inherent in their sex and the environmental response that might be better termed "gender differences." Clinicians have anecdotally agreed for years that patellofemoral (PF) disorders are more common in women. Given the difficulty in classifying patellofemoral disorders, literature support for this assumption is meager [1]. For the purposes of this article we divide PF disorders into three categories: PF pain, PF instability, and PF arthritis. Possible sex differences in these disorders are reviewed.

Patellofemoral pain

One longitudinal study performed by Nimon and associates [2] followed a series of 63 girls (average age 15.5 years) initially identified between the years 1974 and 1980. They chose as their population base girls, because they felt that girls more commonly demonstrated anterior knee pain. At an average of 3.8 years' follow-up, with 54 of the 63 original patients responding to a survey, 50% improved during the first 4 years of this study. A follow-up study of 49 of these patients who responded to a 1994 survey (average follow-up, 16 years) showed that an additional 23 improved in the subsequent 12 years.

When these authors performed statistical analysis of various measurements, including radiographic measurements and physical examination features, they could not identify any factors to predict who failed to improve in this anterior knee pain group. Patients with PF pain who did not improve still had no evidence of significant structural disease, however [3]. Although there are several limitations to this early study, it suggests that at least a subset of PF pain is not related to known structural disease, nor is it related to standard physical examination or radiographic features.

Another group of investigators examined various constitutional features in trying to design a prospective study design for identification of anterior knee pain. They included physical examination features and various constitutional features, including sex, age, and body composition, athletic activities, and duration of symptoms. They found that age was the only factor that showed significant predictive value; the younger the patient, the more favorable the outcome [4]. Sex was not a predictive factor in regards to anterior knee pain in this study group.

A more recent study looked at the development of anterior knee pain in an athletic population [5]. This was a 2-year prospective design using students with no prior history of knee complaints or surgery enrolled in physical education classes. The authors looked at various intrinsic factors, including anthropometric variables, motor performance, general joint laxity, lower leg alignment, muscle length and stretch, and PF characteristics on physical examination. Two hundred eighty-two students (151 boys and 131 girls), average age 18.6 years, were available for review. Of the 151 boys, 11 (7%) developed PF pain compared with 13 of 131 girls (10%). Other findings included a shortened quadriceps muscle and a hypermobile patella as risk factors for PF pain.

E-mail address: arend001@umn.edu

The question of whether PF pain is more prevalent in girls is difficult to answer from current epidemiology studies. The association between clinical overload and PF pain is well established [6,7]. The origin of PF pain is unclear and likely multifactorial [8]. There is clear literature support for reducing anterior knee pain with physical therapy rehabilitation techniques, however, most specifically improving strength [9–12]. Despite the success of this established treatment pathway, fundamental questions remain. Specifically, what is the disturbance in core neuromuscular control strategies that either creates pain or is a result of pain, and how does muscular strengthening change this relationship?

Neuromuscular control of a joint has been a strategy much used in orthopaedics in nonsurgical treatment of clinical pain and joint dysfunction. These conditions have included painful shoulder syndromes, PF overuse syndromes, and even known osteoarthritis of joints, including the knee, back, and shoulder. Improving neuromuscular control and strength of the joint seems to clinically improve function and reduce pain. Little is known about the strategy used by the body to restore normal neuromuscular control and whether it is performed at the central or local level.

Neuromuscular differences exist in the lower extremity between men and women [13]. Women typically demonstrate less muscle mass [14–18] and different muscle fiber compositions [15,19,20]. Physical activity has the most direct influence on muscle composition in any one individual [21,22]. It is also known that hormonal influences, particularly testosterone, can increase muscle mass, fiber recruitment, and a proportion of type II fibers [23]. These factors all favor muscle strength increases in men over women.

Dimorphism can be found in specific muscles in the body. Analysis of muscles on this level is in its infancy. We do know that differences in muscle fiber phenotype develop under the influence of testosterone in men, however. This influence leads to difference in muscle fiber phenotype, which is reflected in the contractile properties of muscle fibers [24]. Androgen treatment has been found to change motor neuron firing rates, which suggests that androgen has influence on the muscle fibers themselves and the motor neurons that influence them [25].

Functional stability of a joint is maintained by muscle recruitment, which depends on a neurocontrolled reflex response. The critical components of stabilizing control of any joint depend on the orchestration of various factors, including the recruitment patterns of voluntary muscles, intrinsic stiffness of contracting muscles, sensory proprioception, and reflex response.

Sex difference in joint stability has been looked at mostly in regards to trunk stabilization. Differences in muscle recruitment and coactivation have been observed in functional task as it relates to trunk stabilization [26]. Sex differences in dynamic joint stiffness of the knee also have been observed [27,28]. Muscle strength and muscle fatigue play important roles in the transference of joint load. The relationship of this control to muscle fatigue and how it relates to the appearance of clinical symptoms of pain is not well understood. Whether this has any relationship to the development of long-term consequences of joint dysfunction and pain continues to be studied.

Development likely plays a critical role in an individual's neuromuscular response to activities. Hewett and colleagues [29] found that before puberty, boys and girls have similar jumping and landing strategies. At puberty, however, these strategies change. After the onset of maturation, female athletes landed with more medial motion of the knee and demonstrated a significant difference between the maximum valgus angles of their dominant and nondominant lower extremities. Sex hormones play a large role in determining the size and shape of soft tissue structures that support a joint, particularly the size and action of muscles that support a joint. At puberty these differences become most manifest. Taken collectively, neuromuscular differences in men and women may help explain sex discrepancy in joint injury patterns and joint pain, including PF pain [13].

Central to the discussion of PF pain is understanding whether pain is perceived differently by women on a qualitative basis or whether pain has different biologic implications in female versus male subjects. Women are more sensitive to pain stimuli, less tolerant of pain, and more able to discriminate different pain patterns [30]. These differences may be present because of qualitative differences in neuroprocessing of pain and analgesia [31]. Areas to explore in unraveling possible dimorphic features in the presentation and treatment of PF pain include (1) pain perception and pathways and (2) differences in baseline neuromuscular responses and strategies for changing or improving those responses.

Patellofemoral dislocations

PF dislocation is defined as the traumatic disruption of normal or previously uninjured tracking relationship of the patella to its groove (Fig. 1). Although dated, most of the published literature on acute patellar dislocations has a predominance of men in the study populations [1,32–35]. This factor likely is because these studies date from a time when men constituted most of the athletes; these differences likely reflect variability in demographics and the entry criteria of the investigator's practice. One population-based study of first-time PF dislocators found an equal number of men and women [36]. A few studies reported that recurrent patellar dislocations occur more frequently in women [37,38].

This information suggests that acute patellar dislocations occur more frequently in men, whereas women may experience more recurrent dislocations. This factor is further borne out by a well-conducted prospective, population-based study [39]. These authors prospectively followed 189 patients for 2 to 5 years and recorded several history, physical examination, and radiographic measurements. The risk of a PF dislocation was highest among girls aged 10 to 17 years old. Patella dislocators who present with a history of PF instability are more likely to be of the female gender.

A prospective study conducted from 1990 to 1993 looked at 177 knees in 133 patients with documented patellar dislocations. Of the 133 patients, 92 (69%) were women [40]. Further study in this area to document the degree of soft tissue and cartilage damage, physical examination features, and treatment rendered would be necessary to determine whether sex is an individual risk factor for recurrent patellar dislocations.

Constitutionally, several physical examination features have a greater preponderance in females. (1) Women have a greater Q angle [41–43]. (2) Women have a higher prevalence of increased femoral anteversion [44–47]. (3) Patella alta is more common in women [40,48,49]. PF instability has been associated with various anatomic features. Patella alta is associated with PF instability [1,36,50–54]. In knees with patella alta, the kneecaps are higher or more cephalad, and it takes a greater degree of flexion before the kneecap is stabilized in the confines of the bony architect of the trochlear groove.

Trochlear dysplasia long has been recognized as a factor in PF instability, first introduced by Albee in 1915, when he proposed a superolateral trochleoplasty [55]. This dysplasia was further analyzed on radiographic images using the axial view [56] and the true lateral radiograph [57]. In 1994, Dejour and colleagues [40] first reported their findings in the English literature. The authors analyzed anatomic factors of instability using radiographic imaging. Their study group was composed of 143 knees (59% female) with symptomatic patellar instability, 67 contralateral asymptomatic knees, and 190 control knees radiographs and 27 control knees CT scans.

Four factors were relevant in knees with symptomatic patellar instability:

1. Trochlear dysplasia (85%), as defined by the crossing sign (96%) (Fig. 2) and the presence of a trochlear bump (66%)
2. Quadriceps dysplasia (83%), which was defined as lateral patellar tilt on CT scan
3. Patella alta (24%)
4. Tibial tuberostiy-trochlear groove distance (56%) defined as a measurement of distance between the two sites on CT scan

The literature suggests that women have a higher rate of certain dysplastic features in anatomy and bony morphology that are individual risk factors for patellar instability. Although there remains debate about sex disparity in first-time PF dislocations, women tend to have more recurrent dislocations. This is likely caused by an increase in certain dysplastic features of the PF joint that are present to a greater degree in women than men.

Treatments for patellar dislocations include improving strength and body awareness of the knee and limb and focusing on rotational control

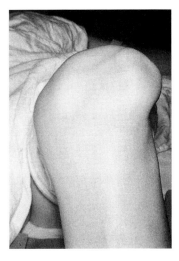

Fig. 1. Typical traumatic patellar dislocations occur with the kneecap translating lateral as the knee flexes.

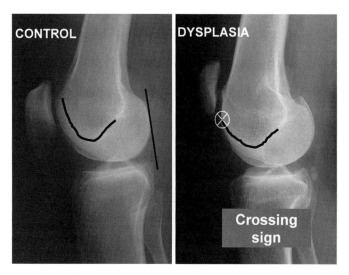

Fig. 2. Lateral radiograph of a standing knee with both condyles on the same line. In knees without trochlear dysplasia, the line of the trochlear groove does not intersect the cortex. In dysplastic knees, the trochlear line crosses the cortex (ie, "crossing sign") representing a shallow and shorter trochlear groove. (Radiographs courtesy of David DeJour MD, Lyon, France.)

of the limb under the pelvis. These same strategies are likely useful for intervening to prevent injury. There is increasing literature support for altered or abnormal neuromuscular control of the lower limb, particularly the knee joint, contributing to the female anterior cruciate ligament injury mechanism [58–61]. Neuromuscular training corrected jump and landing techniques and significantly reduced abduction moments at the knee, and it decreased anterior cruciate ligament injuries in a female intervention group to a rate similar to that of men [62]. Neuromuscular control and morphology of the PF joint likely combines to affect the susceptibility of injury to the PF joint. Understanding the specific neuromuscular risk factor and its possible gender predilection is currently intensely studied in regards to the sexual differences in the incidence of noncontact anterior cruciate ligament injury, (see the article by Slauterbeck elsewhere in this issue). With regards to establishing a risk equation for PF instability, joint morphology is clearly a strong risk factor for injury.

Patellofemoral arthritis

In a 1992 study published in *Annals of Rheumatologic Disease*, McAlindon and colleagues [63] looked at the incidence of PF arthritis judged by radiographs. The study contained 273 subjects with complaints of knee pain and 240 controls. Within this population, isolated PF arthritis was present more than twice as often in women as in men (24% of the women versus 11% of the men). The incidence of medial compartment and PF arthritis showed equal variance between men (7%) and women (6%).

A recent retrospective review of isolated PF arthritis (Fig. 3), which was a multicenter study [49] that involved several French orthopedic centers, supports this greater incidence in women. Of the 578 patients, 72% were women. The main purpose of this study was to review predisposing factors and treatment outcomes. The goal

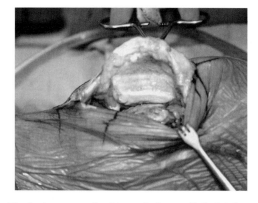

Fig. 3. An open arthrotomy of a knee with isolated patellofemoral arthritis. The patella was mal-tracking on the lateral condyle of the trochlear groove. The patella shows a wide lateral facet, concave in shape, fitting the condylar shape. Eburnated bone is present.

was to correlate predisposing factors in the hopes of better defining the origin of PF arthritis and gain insight into the natural history of this disease. The results divided etiologies into four subgroups: 49% idiopathic (unknown), 33% with objective patellar instability signs, 9% with previous history of blunt trauma, and 8% with chondral calcinosis. In this population, 33% had radiographic evidence of one or more dysplastic features of the PF joint, which are more prevalent in women.

Summary

PF injury and disease are difficult to define and characterize. Frequently, disease classifications are lumped together, which makes it hard to establish meaningful baseline data from our current literature. Population demographics that define PF disorders by sex are scant. Clinical data support the fact that these problems are more common in women. Potential risk factors for PF disease and injury that may show a variance by sex include anatomic and neuromuscular factors. As we better classify PF disorders and tease out potential risk factors, a female sex predilection is likely to become more apparent.

References

[1] Arendt EA. Patellofemoral disorders. In: Garrett WE, Lester G, McGowan J, et al, editors. Women's health in sports and exercise. Rosemont (IL): American Academy of Orthopaedic Surgeons; 2001. p. 125–37.

[2] Nimon G, Murray D, Sandow M, et al. Natural history of anterior knee pain: a 14–20 year follow-up of nonoperative management. J Pediatr Orthop 1998; 18:118–22.

[3] Sandow MJ, Goodfellow JW. The natural history of anterior knee pain in adolescents. J Bone Joint Surg Br 1985;67:36–8.

[4] Kannus P, Niittymaki S. Which factors predict outcome in the nonoperative treatment of patellofemoral pain syndrome? A prospective follow-up study. Med Sci Sports Exerc 1994;26:289–96.

[5] Witvrouw E, Lysens R, Bellemans J, et al. Intrinsic risk factors for the development of anterior knee pain in an athletic population: a two-year prospective study. Am J Sports Med 2000;28:480–9.

[6] Fairbank JC, Pynsent PB, van Poortvliet JA, et al. Mechanical factors in the incidence of knee pain in adolescents and young adults. J Bone Joint Surg Br 1984;66:685–93.

[7] Milgrom C, Finestone A, Eldad A, et al. Patellofemoral pain caused by overactivity: a prospective study of risk factors in infantry recruits. J Bone Joint Surg Am 1991;73:1041–3.

[8] Dye SF. Patellofemoral pain without malalignment: a tissue homeostasis perspective. In: Fulkerson JP, editor. Monograph series. Rosemont (IL): American Academy of Orthopaedic Surgeons; 2005. p. 1–9.

[9] Wilk KE, Davies GJ, Mangine RE, et al. Patellofemoral disorders: a classification system and clinical guidelines for nonoperative rehabilitation. J Orthop Sports Phys Ther 1998;28:307–22.

[10] Witvrouw E, Danneels L, Van Tiggelen D, et al. Open versus closed kinetic chain exercises in patellofemoral pain: a 5-year prospective randomized study. Am J Sports Med 2004;32:1122–30.

[11] McConnell J. The management of chrondromalacia patellae: a long-term solution. Aust J Physiother 1986;32:215–23.

[12] O'Neill DB, Micheli LJ, Warner JP. Patellofemoral stress: a prospective analysis of exercise treatment in adolescents and adults. Am J Sports Med 1992; 20:151–6.

[13] Huston LJ, Wojtys EM. Neuromuscular differences between male and female athletes contributing to anterior cruciate ligament injuries. In: Garrett WE, Lester GE, McGowan J, et al, editors. Women's health in sports and exercise. Rosemont (IL): American Academy of Orthopaedic Surgeons; 2001. p. 347–56.

[14] Kanehisa H, Okuyama H, Ikegawa S, et al. Sex difference in force generation capacity during repeated maximal knee extensions. Eur J Appl Physiol 1996; 73:557–62.

[15] Miller AE, MacDougall JD, Tarnopolsky MA, et al. Gender differences in strength and muscle fiber characteristics. Eur J Appl Physiol 1993;66: 254–62.

[16] Henriksson-Larsen K. Distribution, number and size of different types of fibres in whole cross-sections of female m tibialis anterior: an enzyme histochemical study. Acta Physiol Scand 1985;123: 229–35.

[17] Ricoy JR, Encinas AR, Cabello A, et al. Histochemical study of the vastus lateralis muscle fibre types of athletes. J Physiol Biochem 1998;54:41–7.

[18] Sale DG, MacDougall JD, Alway SE, et al. Voluntary strength and muscle characteristics in untrained men and women and male bodybuilders. J Appl Physiol 1987;62:1786–93.

[19] Nygaard E. Skeletal muscle fibre characteristics in young women. Acta Physiol Scand 1981;112: 299–304.

[20] Simoneau JA, Bouchard C. Human variation in skeletal muscle fiber-type proportion and enzyme activities. Am J Physiol 1989;257:E567–72.

[21] Hakkinen K, Komi PV. Changes in neuromuscular performance in voluntary and reflex contraction during strength training in man. Int J Sports Med 1983;4:282–8.

[22] Zappala A. Influence of training and sex on the isolation and control of single motor units. Am J Phys Med 1970;49:348–61.

[23] Ramos E, Frontera WR, Llopart A, et al. Muscle strength and hormonal levels in adolescents: gender related differences. Int J Sports Med 1998;19:526–31.

[24] English AW, Schwartz G. Development of sex differences in the rabbit masseter muscle is not restricted to a critical period. J Appl Physiol 2002;92:1214–22.

[25] Tosi LL, Boyan BD, Boskey AL. Does sex matter in musculoskeletal health? J Bone Joint Surg Am 2005;87:1631–47.

[26] Granata KP, Wilson SE, Padua DA. Gender differences in active musculoskeletal stiffness. Part 1. Quantification in controlled measurements of knee joint dynamics. J Electromyogr Kinesiol 2002;12:119–26.

[27] Granata KP, Padua DA, Wilson SE. Gender differences in active musculoskeletal stiffness. Part II. Quantification of leg stiffness during functional hopping tasks. J Electromyogr Kinesiol 2002;12:127–35.

[28] Blackburn JT, Riemann BL, Padua DA, et al. Sex comparison of extensibility, passive, and active stiffness of the knee flexors. Clin Biomech (Bristol, Avon) 2004;19:36–43.

[29] Hewett TE, Myer GD, Ford KR. Decrease in neuromuscular control about the knee with maturation in female athletes. J Bone Joint Surg Am 2004;86:1601–8.

[30] Fillingim RB, Ness TJ. Sex-related hormonal influences on pain and analgesic responses. Neurosci Biobehav Rev 2000;24:485–501.

[31] Mogil JS, Wilson SG, Chesler EJ, et al. The melanocortin-1 receptor gene mediates female-specific mechanisms of analgesia in mice and humans. Proc Natl Acad Sci U S A 2003;100:4867–72.

[32] Hawkins RJ, Bell RH, Anisette G. Acute patellar dislocations: the natural history. Am J Sports Med 1986;14:117–20.

[33] Cash JD, Hughston JC. Treatment of acute patellar dislocation. Am J Sports Med 1988;16:244–9.

[34] Cofield RH, Bryan RS. Acute dislocation of the patella: results of conservative treatment. J Trauma 1977;17:526–31.

[35] Vanioppa S, Laasonon E, Silvennionen T, et al. Acute dislocation of the patella: a prospective review of operative treatment. J Bone Joint Surg Am 1990;72:365–9.

[36] Atkin DM, Fithian DC, Marangi KS, et al. Characteristics of patients with primary acute lateral patellar dislocation and their recovery within the first 6 months of injury. Am J Sports Med 2000;28:472–9.

[37] Halbrecht JL. Acute dislocation of the patella. In: Fox JM, Del Pizzo W, editors. The patellofemoral joint. New York: McGraw-Hill; 1993. p. 123–56.

[38] Larsen E, Lauridsen F. Conservative treatment of patella dislocations: influence of evident factors on the tendency to redislocate and the therapeutic result. Clin Orthop 1982;171:131–6.

[39] Fithian DC, Paxton EW, Stone ML, et al. Epidemiology and natural history of acute patellar dislocation. Am J Sports Med 2004;32:1114–21.

[40] Dejour H, Walch G, Nove-Josserand L, et al. Factors of patellar instability: an anatomic radiographic study. Knee Surg Sports Traumatol Arthrosc 1994;2:19–26.

[41] Horton MG, Hall TL. Quadriceps femoris muscle angle: normal values and relationships with gender and selected skeletal measures. Phys Ther 1989;69:897–901.

[42] Hsu RW, Himeno S, Coventry MB, et al. Normal axial alignment of the lower extremity and load-bearing distribution at the knee. Clin Orthop 1990;255:215–27.

[43] Woodland LH, Francis RS. Parameters and comparisons of the quadriceps angle of college-aged men and women in the supine and standing positions. Am J Sports Med 1992;20:208–11.

[44] Staheli LT. Rotational problems of the lower extremities. Orthop Clin North Am 1987;18:503–12.

[45] Hvid I, Andersen LI. The quadriceps angle and its relation to femoral torsion. Acta Orthop Scand 1982;53:577–9.

[46] Braten M, Terjesen T, Rossvoll I. Femoral anteversion in normal adults: ultrasound measurements in 50 men and 50 women. Acta Orthop Scand 1992;63:29–32.

[47] Prasad R, Vettivel S, Isaac B, et al. Angle of torsion of the femur and its correlates. Clin Anat 1996;9:109–17.

[48] Ahlback S, Mattsson S. Patella alta and gonarthrosis. Acta Radiol Diagn (Stockh) 1978;19:578–84.

[49] Dejour D, Allain J. Symposium about isolated patellofemoral arthritis. Rev Chir Orthop Reparatrice Appar Mot 2004;1(Suppl):S69–129.

[50] Insall J, Goldberg V, Salvati E. Recurrent dislocation and the high-riding patella. Clin Orthop 1972;88:67–9.

[51] Aglietti P, Insall JN, Cerulli G. Patellar pain and incongruence: Part 1. Measurements of incongruence. Clin Orthop 1983;176:217–24.

[52] Blackburne JS, Peel TE. A new method of measuring patellar height. J Bone Joint Surg Br 1977;59:241–2.

[53] Lancourt JE, Cristini JA. Patella alta and patella infera: their etiological role in patellar dislocation, chondromalacia, and apophysitis of the tibial tubercle. J Bone Joint Surg Am 1975;57:1112–5.

[54] Nagamine R, Otani T, White SE, et al. Patellar tracking measurement in the normal knee. J Orthop Res 1995;13:115–22.

[55] Albee FH. The bone graft wedge in the treatment of habitual dislocation of the patella. The Medical Record 1915;88:257–9.

[56] Brattstrom H. Shape of the intercondylar groove normally and in recurrent dislocation of patella: a clinical and x-ray anatomic investigation. Acta Orthop Scand 1964;68(Suppl):1.

[57] Maldague B, Malghem J. Apport du cliche de profil du genou dans le depistage des instabilites rotuliennes: rapport preliminaire. Rev Chir Orthop 1985;71(Suppl II):5–13.

[58] Hewett TE. Neuromuscular and hormonal factors associated with knee injuries in female athletes: strategies for intervention. Sports Med 2000;29:313–27.

[59] Hewett TE, Paterno MV, Myer GD. Strategies for enhancing proprioception and neuromuscular control of the knee. Clin Orthop 2002;402:76–94.

[60] Lloyd DG. Rationale for training programs to reduce anterior cruciate ligament injuries in Australian football. J Orthop Sports Phys Ther 2001;31:645–54 [discussion: 61].

[61] McLean SG, Lipfert S, van den Bogert AJ. Effect of gender and defensive opponent on the biomechanics of sidestep cutting. Med Sci Sports Exerc 2004;36:1008–16.

[62] Hewett TE, Lindenfeld TN, Riccobene JV, et al. The effect of neuromuscular training on the incidence of knee injury in female athletes: a prospective study. Am J Sports Med 1999;27:699–705.

[63] McAlindon TE, Snow SW, Cooper C, et al. Radiographic patterns of osteoarthritis of the knee joint in the community: the importance of the patellofemoral joint. Ann Rheum Dis 1992;51:844–9.

Osteoporosis: Differences and Similarities in Male and Female Patients

Joseph Michael Lane, MD[a,b,c,*], Alana Carey Serota, MD, CCFP[c], Bradley Raphael, MD[c]

[a]Department of Orthopedic Surgery, Weill Medical College of Cornell University,
1300 York Avenue, New York 10021, NY
[b]Metabolic Bone Disease Service, Hospital for Special Surgery,
535 East 70th Street, New York, NY 10021, USA
[c]Department of Orthopedics, Hospital for Special Surgery, 535 East 70th Street, New York, NY 10021, USA

Osteoporosis is a widespread disorder. It is a metabolic bone disease that affects more than 200 million individuals worldwide, making it one of the most common global health issues. It is characterized by low bone mineral density (BMD) and a deterioration of bony architecture leading to fragility and fracture. BMD is measured by dual energy x-ray absorptiometry (DXA) and reported as a T-score comparing the subject to young healthy individuals at peak bone mass. The World Health Organization has defined osteoporosis as a T-score ≥2.5 standard deviations below the mean for young normals [1].

Epidemiology

In the United States alone, 10 million individuals are estimated to have osteoporosis, and the condition affects 45% of women over the age of 50. Osteoporosis affects a smaller percentage of men—on the order of 16%—and occurs approximately one decade later than women [2–4].

When BMD decreases, the risk of fracture increases [5]. As the United States faces a "senior boom," these numbers will continue to rise, yet the disease continues to go undetected and undertreated in many cases. Undertreatment occurs partly because most individuals are not diagnosed until fractures occur, causing significant morbidity and mortality.

More than 1.5 million low-energy fragility fractures occur in the United States per year: 700,000 in the spine, 300,000 in the hip, and 200,000 in the wrist (Fig. 1) [6]. At 50 years of age, the total lifetime risk for a fracture of the hip, spine, or wrist is almost 40% for women and 13% for men (Table 1) [7]. Estimates show that the number of hip fractures worldwide will grow from 1.26 million in 1990 to 2.6 million by 2025 and to 4.5 million by the year 2050 [8]. Hip fractures cause the most significant morbidity and mortality among all the effects of osteoporosis, including pain, immobility, depression, and death.

Vertebral fracture rates are relatively comparable between men and women between the ages of 55 and 59. With increasing age, however, women have a 2:1 ratio of vertebral fractures, as reported by Cummings and colleagues [9]. By the age of 75, 13.6% of men and 29.3% of women have vertebral fractures (Fig. 2). Similarly, with respect to hip fractures, men experience these fractures at approximately one third the rate of women. Chang and colleagues [10] reported that approximately one half of hip fractures occurred before the age of 80 in men and two thirds occurred before age 85 in women. In the United States and Northern Europe, approximately 20% to 25% of hip fractures occur in men [11]. The overall mortality rate for hip fractures is 20%; however, men have more than twofold the

* Corresponding author.
E-mail address: lanej@hss.edu (J.M. Lane).

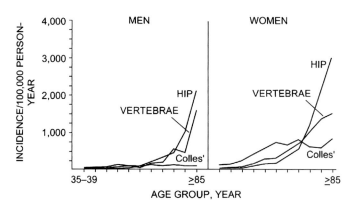

Fig. 1. Age-related fractures. (*From* Cooper C, Melton LJ. Epidemiology of osteoporosis. Trends Endocrinol Metab 1992;3:224–9; with permission.)

mortality rate for women and hip fractures at any given age. The number of men who succumb to hip fractures is similar to the number of women who succumb to their fractures, although the number of women who fracture a hip is much greater than the number of men with the same injury.

Fracture risk increases with age and with decreased bone density. Although the relative risk of fracture increases with decreasing bone density, however, most patients with hip fractures clearly fall within the osteopenic range. For example, Chang and colleagues [10] found that 40% of women aged 65 or older with a hip fracture only have osteopenia, not osteoporosis. Similarly, 74% of those women with nonvertebral fractures had T-scores better than −2.5. A similar relationship occurs among men with hip fractures. This finding indicates that issues beyond decreased bone density account for the increased fractures. Similarly, geographic variation in hip fracture incidence strongly indicates that the risk of fracture is multifactorial (Fig. 3).

Table 1
Estimated lifetime risk of fracture in 50-year-old white men and women

Fracture site	Men (%)	Women (%)
Hip fracture	6.0	17.5
Clinically diagnosed vertebral fracture	5.0	15.6
Distal forearm fracture	2.5	16.0
Any of the above	13.1	39.7

Data from Melton LJ 3rd, Chrischilles EA, Cooper C, et al. Perspective. How many women have osteoporosis? J Bone Miner Res 1992;7:1005–10.

Anatomy and pathophysiology

Osteoporosis is related to decreased bone mass, inadequate distribution of the bone mass, a loss of microarchitectural structure, inconnectivity, and a failure of the bone modulus. This last factor relates to material properties of the bone, which in this case is determined by the mineral content and the collagen quality and quantity. Mineral provides the compressive strength involved and can be altered by such diseases as osteomalacia, which involves a failure of mineralization. The tensile strength of bone is related to the collagen, which is most evident in disorders such as osteogenesis imperfecta, which involves a failure of collagen form, shape, and structure. Men and women reach peak bone density in their

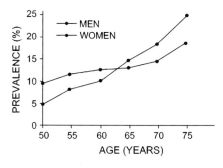

Fig. 2. Prevalence of vertebral deformity by sex. Data derived from the European Vertebral Osteoporosis Survey. (*From* O'Neill TW, Felsenberg D, Varlow J, et al. The prevalence of vertebral deformity in European men and women: the European Vertebral Osteoporosis Study. J Bone Miner Res 1996;11(7):1011–9; with permission of the American Society for Bone and Mineral Research.)

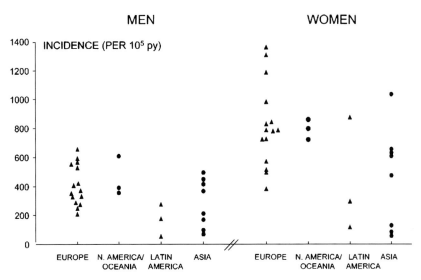

Fig. 3. Geographic variation in hip fracture incidence. (*Data from* Kanis JA, Johnell O, de Laet C, et al. International variations in hip fracture probabilities: implications for risk assessment. J Bone Miner Res 2002;17:1237–44.)

early 20s, somewhat later in the men than women [12]. Peak bone mass is, on average, 10% to 12% greater in men than in women. Although the amount and thickness of trabecular bone is similar between the sexes, men have thicker cortical bone and their bones are larger, even when body size is taken into account. It is postulated that this difference accounts for the decreased risk of fracture in men and the delayed onset of the disease.

Osteoporosis is generally characterized into primary and secondary causes. Primary osteoporosis is divided into type I, postmenopausal osteoporosis, and type II, senile or age-related osteoporosis. Type I represents a high turnover state that occurs after menopause. Type II is largely a failure of osteoblastic bone to form. Up to 90% of peak bone mass is acquired by age 18 in girls. After menopause, properly defined as the last menstrual period, there is a precipitous drop of bone by approximately 2% per year for approximately 6 years, which is associated with a high turnover state. The rate of bone loss then decreases to a slow attrition with time [13]. In contradistinction to women, once men achieve peak bone mass at a slightly later time than women, they stay steady for a period of time and then have a slow attrition without a period of rapid turnover [14]. In men who develop acute hypogonadism at any age, however, bone loss accelerates to a rate similar to that of menopausal women.

The bone loss that follows orchiectomy for prostate cancer, for example, is rapid for several years and then reverts to the aforementioned, age-associated gradual loss [15].

Secondary osteoporosis is caused by a multitude of factors, including endocrine disorders, hematopoietic diseases, immobilization, gastric disorders, and medications. Among men, more than 30% to 60% of osteoporosis is considered secondary, the most common causes of which are increased alcohol intake and smoking [16]. Long-term oral glucocorticoid therapy accounts for nearly one in six cases of male osteoporosis [17]. Conversely, in women osteoporosis is usually primary, type I.

Risk factors and prevention

Cummings and co-investigators [5] identified a series of risk factors for hip fractures, including low bone mass, increasing age, history of falls, previous fractures, certain ethnic groups (eg, white), female gender, small body size, high bone turnover, dimensions of the bone, bone geometry, frailty of the individual, sedative medicines, use of anticonvulsants, drugs for hypertension, polypharmacy, vitamin A toxicity, vitamin D deficiency, calcium deficiency, protein deficiency, cerebrovascular accidents, poor vision, hyperthyroidism, family history of fractures, peripheral neuropathy, type I diabetes, dementia, Parkinson's disease, and history of recent weight loss. These factors seem to increase the risk individually or in combination for fractures and they apply to men and women. Because women live longer than men, they are more likely to have one or more of these risk factors affect their lives. The greatest risk factor for additional fracture is having a prior fragility fracture [16]. For example, the presence of one or more vertebral fractures is a risk factor, and it increases the risk of a subsequent vertebral fracture fivefold and a subsequent hip fracture twofold [18]. These observations are found in men and women. Postmenopausal women are most at risk for fracture as a consequence of the loss of the protective effects of estrogen. The overall lifetime risk of hip fracture for women is 14%, which increases dramatically with age. Lin and Lane [19] reported that 20% of women will have a hip fracture by age 80. This number increases to 50% by the age of 90, and women older than 85 are approximately eight times more likely than women aged 65 to 74 to be admitted to the hospital for a hip fracture.

There has been increased investigation into the bone protective effects of estrogen for women and testosterone for men and the benefits of early DXA screening. Van der Klift and colleagues [20] described in their cohort study that overall incidence of vertebral fractures is higher in women, but the correlation between BMD and vertebral fracture in men is similar to the relationship between BMD and fractures in women. For both genders, the incidence of vertebral fractures doubled per standard deviation decrease in lumbar spine or femoral neck BMD. When looking at gender differences in hip fractures among cohorts, de Laet and colleagues [21] showed that average hip fracture risk in men is much lower than in women but seemed to be similar at the same BMD. They proposed using the same absolute BMD thresholds for decisions about intervention. Some studies have suggested that the relationship between BMD and hip fracture risk is different in men and women [22,23].

The risk of hip fracture increases with the height of an individual and decreased body mass. Men are generally taller than women but have greater body mass; it is more difficult to strike the greater trochanter in men than in thin women. Balance is critical, and exercise aimed at improving balance is important. A proven intervention is the martial art Tai-Chi, which can decrease falls by approximately 50% [24]. Early data on hip protectors were promising, but recent studies suggest that they are ineffective for individuals who live at home and are of questionable value in individuals in institutional settings [25].

Diagnosis

The diagnosis of osteoporosis is primarily made through a bone density instrumentation, which determines an aerial density mineral content versus a three-dimensional volumetric density. In the DXA machine, measurements are taken of the lumbar spine (primarily L1, 2, 3, and 4) and the hip, notably the femoral neck and total femur. The mineral content is then corrected for the aerial dimensions and compared with gender-specific standards. Although there has been an effort to set up standards for the different ethnic groups, it seems that the standards for white men and women apply well to groups across ethnic boundaries. The Z-score is the value that compares a patient's bone density and that of his or her peers, and the T-score compares one to an idealized peak bone mass of young normal. Values that are up to one standard deviation below the mean for young, healthy subjects are considered normal; values less than -1.0 to -2.5 standard deviations are considered an indication of osteopenia, and values -2.5 or less are considered an indication of osteoporosis [1]. These numbers were created for epidemiologic and demographic studies and do not necessarily reflect clinical practice thresholds for intervention.

The lowest T-score value of the aforementioned sites is used to determine an individual's status. The instrument precision for the spine in young individuals is 2% and that of the hip is closer to 3% [26]. In individuals past the age of 60 who have scoliosis, osteophytes, or other abnormalities, the precision falls off in the spine and clinical attention is usually redirected toward the hip. The total femur has better precision than the femoral neck, and evaluation of serial measurements are best suited to that area rather than the neck, although the latter may present the lowest bone density. The most comprehensive list of indications for performing a DXA is published by the International Society for Clinical Densitometry. It includes women aged 65 years and older, postmenopausal women under age 65 years with risk factors for osteoporosis, men aged 70 years and older, adults with fragility fracture, adults with a disease or condition who take medication associated with low bone mass or bone loss, anyone being considered for pharmacologic osteoporosis therapy, anyone being treated for low bone mass to monitor treatment effect, anyone not receiving therapy in whom evidence of bone loss would lead to treatment, and women discontinuing estrogen according to the indications listed above [50].

The evaluation of a patient with an abnormal DXA or a fragility fracture should be directed at four primary areas, including abnormalities of the hematopoietic system (eg, multiple myeloma), endocrinopathies (including hyperthyroidism and hyperparathyroidism), nutritional deficiencies (including hypovitaminosis D), and renal function. Initial screening tests should include a complete blood count, calcium, bone-specific alkaline phosphatase, kidney and liver function tests, 25-hydroxyvitamin D, intact parathyroid hormone (PTH), and thyroid-stimulating hormone level. Depending on the history and the uniqueness of the individual, further studies would be warranted. Total testosterone may be measured in men and an estradiol level may be measured in women and men. Although a low estradiol level in men may contribute to osteoporosis, no accepted treatment is available [27]. Women who develop osteoporosis are often small, have limited body fat, and have been shown to have low levels of estradiol, but there is no clear correlation between estrogen levels and osteoporosis. Among men, testosterone levels decrease with age but it is rarely in the pathologic range. Testosterone replacement increases the risk of prostate cancer and cardiac disease and is rarely used unless men are manifesting other symptoms of deficiency, including sexual dysfunction, depression, low energy, and decreased muscle mass.

The final area of interest is the determination of high versus low bone turnover. Bone formation rates are measured with bone-specific alkaline phosphatase, osteocalcin, and amino-terminal propeptide of type 1 collagen, although the latter is not yet commercially available. Markers for increased bone resorption are determined by collagen breakdown products and can be measured in the urine with pyridinoline, deoxypyridinoline, N-telopeptide crosslinked collagen type 1, and serum levels of C-telopeptide crosslinked collagen type 1. Because most agents currently in use for osteoporosis are antiresorptive agents, the N-telopeptide crosslinked collagen type 1 and the C-telopeptide crosslinked collagen type 1 are the most frequently evaluated levels. In high turnover osteoporosis there is increased osteoclastic resorption, and the osteoblasts are unable to keep up with the depth and the extent of the osteoclastic stages. These changes are recognized by high values of collagen breakdown products. Conversely, in low turnover osteoporosis in which there is a failure of osteoblastic bone formation, the N-telopeptide crosslinked collagen type 1 may be normal or low, and the markers for bone formation are particularly suppressed [28]. The evaluations for men and women are relatively similar, with the expectation that in men either by history or by laboratory studies there will be a higher rate of secondary causes of osteoporosis. Beyond bone density determination and laboratory findings are individual risk factors that determine the way in which treatment is established. These prominent risk factors as recognized by the National Osteoporosis Foundation include low body weight, recent loss in body weight, history of low-energy fracture after the age of 50, history of a fragility fracture in a first-degree relative (most notably the maternal or paternal), and a history of active smoking. The presence of these factors would lower the threshold for initiating therapy. Weight deficiency is most commonly found among women, whereas the rate of smoking is clearly higher among men.

Treatment

The therapeutic approach to osteoporosis has been enhanced with the development of a whole family of antiresorptive agents and, recently, the introduction of anabolic agents. Any and all treatment begins with adequate calcium and vitamin D, which applies to women and men. In several reviews of existing research, there is a clear dichotomy in the efficacy for calcium and vitamin D. In studies in which the dose of vitamin D was ≤ 400 IU daily there seems to be no benefit with respect to fracture prevention. In studies in which calcium was given with ≥ 800 IU daily of vitamin D, the risk of fracture decreased, particularly in elderly and debilitated individuals. In nursing homes, the data from Chapuy and colleagues [29] suggested that hip fracture rates can be cut by 43%. Osteomalacia seems to be widespread in older adults, who benefit most from the extra vitamin D and calcium. Studies have indicated that one has to supplement 800 IU daily to maximize muscle function, and there seems to be no preference of women over men for osteomalacia. Recent studies among patients with hip fractures suggest that as many as 60% to 70% of individuals may be vitamin D deficient, defined as a 25 hydroxyvitamin D level of <80 nM [30].

Calcium can be given as a carbonate and as a citrate. Although the carbonate contains a greater amount of elemental calcium per pill, its absorption depends on a low pH and it use can be associated with an increased risk of kidney stones, as was demonstrated in the Women's Health Initiative [31]. Citrate, on the other hand, is dissolved in any pH, so is useful in older patients and patients who are taking H2-blockers and proton pump inhibitors. It also can bind to the oxylate within the gastrointestinal tract and lower the chance of kidney stones. The preferred form of vitamin D is vitamin D3, cholecalciferol. In patients who have osteoporosis, the recommended daily intake of calcium is between 1200 and 1500 mg, with at least 800 IU of vitamin D. Men have a greater risk of kidney stones, so if there is a history of same, the urinary calcium should be kept under 100 mg/L.

Estrogens, both natural and synthetic, have been used to treat osteoporosis in women. Many studies, including the Women's Health Initiative, have demonstrated clearly that estrogen replacement can increase bone mass approximately 2% per year and decrease the risk of hip and spine fracture by approximately 35% [32]. In the Women's Health Initiative, women on combined conjugated equine estrogen and medroxyprogesterone acetate had an increased risk of myocardial infarction, deep venous thrombosis, cerebrovascular accident, and dementia and a 2% per year increased risk of developing breast cancer over the rate for their peers. Participants in the estrogen-only arm showed an increased risk of cerebrovascular accident and deep venous thrombosis [33]. As a consequence, estrogen is no longer considered an appropriate treatment for postmenopausal osteoporosis and should be used only in patients who experience symptoms of estrogen deficiency [32]. The selective estrogen receptor modulators, namely tamoxifen and raloxifene, exert antiestrogenic effects on breast tissue and estrogenic effects on the bone. Both of these agents can prevent bone loss, but tamoxifen cannot be used for the treatment of osteoporosis because of the high rate of uterine cancer and severe symptoms of estrogen deficiency, specifically hot flashes and vaginal dryness. Raloxifene has been demonstrated to increase the bone mass of the spine and decrease the risk of vertebral fractures on the order of 35% [34]. In a large series, however, it did not show even a trend toward preventing hip fractures. It is also associated with profound postmenopausal symptomatology, particularly if therapy is initiated within 5 years of natural menopause, but it does not increase the risk of uterine cancer. Recent analysis of the data from the study of tamoxifen and raloxifene (STAR) trial demonstrated that raloxifine decreases the risk of invasive breast cancer and has a lower rate of deep venous thrombosis when compared with tamoxifen (news released on NCI website on April 27, 2006). Although there was initial hope that these agents would prove useful in men, it seems that men metabolize the drug before its biologic activity can be exerted. Some men are testosterone deficient, but replacement is used only rarely in them because of the possible increased risk of prostate cancer and heart disease.

Bisphosphonates are pyrophosphonate analogs that are nondegradable and bind firmly to bone. They form a shield that precludes the osteoclastic resorption of bone; if absorbed by the osteoclasts, they lead to the premature death of the cell caused by inhibition of membrane synthesis and alteration of the Krebs cycle. Agents such as alendronate, risedronate, and ibandronate have gone through extensive testing and have been shown to be effective in increasing bone density of all bones and decreasing fractures of all bones by up to approximately 50% [35]. They are all highly acidic and can cause indigestion. Alendronate and risedronate are available as weekly preparations, and ibandronate is available as a monthly pill or an injection that is given every 3 months. These drugs were originally targeted and evaluated in women and have been shown to be most effective in that population. Recent findings have confirmed that they work comparably in men [36]. As a consequence of the smaller power in the male studies, the efficacy for specific fracture prevention has been difficult to prove, but—particularly with respect to vertebral fractures—these agents seem to be equally efficacious in men.

With time, the bisphosphonates decrease bone turnover, and at high levels in animals they decreased bone strength and resilience. There are some limited reports of patients on long-term bisphosphonates developing transverse stress fractures. Biopsies of these individuals have suggested extremely low turnover states [37]. Although the bisphosphonates are outstanding in their efficacy, bone turnover markers should be monitored. If they become profoundly suppressed, a patient should be taken off the bisphosphonates and given a rest period until the patient can return to therapeutic levels (N-telopeptide crosslinked collagen type 1 20 to 40).

In the area of orthopedics, further efficacy has been demonstrated for the bisphosphonates for preventing the collapse associated with avascular necrosis [38]. Limited studies suggest that they may limit osteolysis and are useful in disuse osteoporosis. In the few evidence-based studies evaluating their effect on fracture healing, it seems that in the metaphyseal and epiphyseal regions the bisphosphonates do not inhibit fracture healing clinically. The only concern is raised with patients who have been on long-term therapy with a low turnover state who develop a transverse fracture [37]. In a diaphyseal fracture it is more prudent to stop the medication for at least 3 to 6 weeks to allow the biology to perform its early stages of regeneration and then restart the bisphosphonates.

The final medication in this category is calcitonin. In postmenopausal women it decreases the risk of spinal fractures by 37% but has no effect on hip fractures [39]. It is not terribly efficacious in terms of overall bone health but has been reported to provide some measure of pain relief [40]. Calcitonin is considered a second-line drug and clearly has none of the efficacy of the bisphosphonates. Few data are available relating to its activity in men, but it seems to have similar benefits and limitations.

Teriparatide, human recombinant parathyroid hormone 1-34, is the only available anabolic agent for the treatment of osteoporosis. When parathyroid hormone is given continuously, it is associated with increased osteoclastic and osteoblastic turnover leading to a net loss of bone. In intermittent subcutaneous administration of 20 µg daily, however, PTH has been demonstrated to lead to an active anabolic phase, with bone mass increasing up to 13% over 2 years in the spine and, to a lesser degree, within the hip [41–43]. Most studies with PTH have been performed on women. The medication decreases the risk of vertebral and nonvertebral fractures to the same extent as the bisphosphonates. Teriparatide is given for a maximum of 2 years, after which time the gains in BMD are sealed in and even augmented with a bisphosphonate, otherwise the BMD drifts down to pretreatment levels [44]. Initial studies using a combination of concurrent PTH and bisphosphonate showed decreased benefit than either agent alone [45]; it is generally recommended that these drugs be given separately and in sequence. A recent study by Cosman and colleagues [46] challenged this conclusion by giving 3 months on/3 months off pulses of teriparatide while the subjects were on weekly alendronate. BMD in the spine increased above that of the alendronate-only arm. It seems that this pulsed regimen takes advantage of the 3- to 4-month so-called "anabolic window" in which the markers of bone formation rise more quickly than the markers of bone resorption. Recent studies in women have shown that the concurrent use of estrogen or raloxifene can enhance the bone-forming effects of teriparatide [47,48]. The use of PTH in men has much more limited data but it seems to have relatively comparable efficacy. Indications for PTH in men and women include individuals who have bone density decline on a bisphosphonate, stabilize on a bisphosphonate at an extremely low level, develop a subsequent fracture while on bisphosphonate therapy, or present initially with a low turnover in which anabolic effect is clearly warranted.

Unique to the female population is the possible use of all the aforementioned agents during the childbearing years. The long half-life of bisphosphonates and their still longer presence in the bone makes one reluctant to use these agents in fertile women. No such concern exists with respect to spermatogenesis. It is tempting to consider PTH in the premenopausal population, with appropriate contraception, because the drug remains in the body for such a short time. As the indications for PTH treatment continue to expand, such as implant integration and fracture healing [49], so too must our consideration of previously unstudied populations.

Summary

Osteoporosis is a disease of fragility fractures related to decreased bone density and poor bone quality. Men and women develop osteoporosis, with the disease in men more often a consequence of secondary causes and occurring at a later period of time. Men have a lower rate of fracture; however, they have a much higher mortality rate than women at any given age. Diagnosis is relatively comparable between men and women and the treatment is similar, with the exception of the selective estrogen receptor modulators. New treatments are currently under development, but our efforts also should be aimed at prevention and early detection.

References

[1] Genant HK, Cooper C, Poor G, et al. Interim report and recommendations of the World Health Organization Task Force for Osteoporosis. Osteoporos Int 2001;10(4):259–64.

[2] Hannan MT, Felson DT, et al. Bone mineral density in elderly men and women: results from the Framingham osteoporosis study. J Bone Miner Res 1997;7(5):547–53.

[3] Jones G, Nguyen T, Sambrook P, et al. Progressive loss of bone in the femoral neck in elderly people: longitudinal findings from the Dubbo osteoporosis epidemiology study. BMJ 1994;309(6956):691–5.

[4] Khosla S, Melton LJ III, et al. Relationship of serum sex steroid levels and bone turnover markers with bone mineral density in men and women: a key role for bioavailable estrogen. J Clin Endocrinol Metab 1998;83(7):2266–74.

[5] Cummings SR, Black DM, et al. Bone density at various sites for prediction of hip fractures: the Study of Osteoporotic Fractures Research Group. Lancet 1993;341(8837):72–5.

[6] Riggs BL, Melton LJ III. The worldwide problem of osteoporosis: insights afforded by epidemiology. Bone 1995;17(5 Suppl):505S–11S.

[7] Melton LJ, Chrischilles EA, Cooper C, et al. How many women have osteoporosis? J Bone Miner Res 2005;7:1005–10.

[8] Gullberg B, Johnell O, et al. World-wide projections for hip fracture. Osteoporos Int 1997;7(5):407–13.

[9] Cummings SR, Karpf DB, Harris F, et al. Improvement in spine bone density and reduction in risk of vertebral factures during treatment with antiresorptive drugs. Am J Med 2002;112(4):281–9.

[10] Chang KP, Center JR, et al. Incidence of hip and other osteoporotic fractures in elderly men and women: Dubbo Osteoporosis Epidemiology Study. J Bone Miner Res 2004;19(4):532–6.

[11] Cooper C, Campion G, et al. Hip fractures in the elderly: a world-wide projection. Osteoporos Int 1992;2(6):285–9.

[12] Gilsanz V. Accumulation of bone mass during childhood and adolescence. In: Orwoll ES, editor. Osteoporosis in men. San Diego (CA): Academic; 1999. p. 65–85.

[13] Delaney MF. Strategies for the prevention and treatment of osteoporosis during early postmenopause. Am J Obstet Gynecol 2006;194(2 Suppl):S12–23.

[14] Seeman E. Sexual dimorphism in skeletal size, density, and strength. J Clin Endocrinol Metab 2001;86:4576–84.

[15] Campion J, Maricic MJ. Osteoporosis in men. Am Fam Physician 2003;67(7):1521–6.

[16] Nguyen TV, Eisman JA, et al. Risk factors for osteoporotic fractures in elderly men. Am J Epidemiol 1996;144(3):255–63.

[17] Seeman E, Melton LJ III, O'Fallon WM, Riggs BL. Risk factors for spinal osteoporosis in men. Am J Med 1983;75:977–83.

[18] Lindsay R, Silverman SL, et al. Risk of new vertebral fracture in the year following a fracture. JAMA 2001;285(3):320–3.

[19] Lin JT, Lane JM. Osteoporosis: a review. Clin Orthop Relat Res 2002;425:126–34.

[20] Van Der Klift M, De Laet CEDH, et al. The incidence of vertebral fractures in men and women: the Rotterdam Study. J Bone Miner Res 2001;17(6):1051–6.

[21] de Laet CE, van der Klift M, et al. Osteoporosis in men and women: a story about bone mineral density thresholds and hip fracture risk. J Bone Miner Res 2002;17(12):2231–6.

[22] Orwoll ES. Assessing bone density in men. J Bone Miner Res 2000;15(10):1867–70.

[23] Melton LJ, Orwoll ES, et al. Does bone density predict fractures comparably in men and women? Osteoporos Int 2001;12(9):707–9.

[24] Gillespie LD, Gillespie WJ, Robertson MC, et al. Interventions for preventing falls in elderly people. Cochrane Database Syst Rev 2003;4:CD000340.

[25] Parker MJ, Gillespie WJ, Gillespie LD. Effectiveness of hip protectors for preventing hip fractures in elderly people: a systematic review. BMJ 2006;332(7541):571–4.

[26] Baim S, Wilson CR, Lewiecki EM, et al. Precision assessment and radiation safety for dual-energy X-ray absorptiometry: position paper of the International Society for Clinical Densitometry. J Clin Densitom 2005;8(4):371–8.

[27] Orwoll ES. Osteoporosis in men. Endocrinol Metab Clin North Am 1998;27:349–67.

[28] Miller PD, Baran DT, Bilezikian JP, et al. Practical clinical application of biochemical markers of bone turnover: consensus of an expert panel. J Clin Densitom 1999;2:323–80.

[29] Chapuy MD, Arlot ME, et al. Vitamin D3 and calcium to prevent hip fractures in the elderly women. N Engl J Med 1992;327(23):1637–42.

[30] Holick MF. High prevalence of vitamin D inadequacy and implications for health. Mayo Clinic Proc 2006;81(3):353–73.

[31] Jackson RD, LaCroix AZ, Gass M, et al. Calcium plus vitamin D supplementation and risk of fractures. N Engl J Med 2006;354(7):669–83.

[32] Rossouw JE, Anderson GL, et al. Risks and benefits of estrogen plus progestin in healthy postmenopausal women: principal results from the Women's Health Initiative randomized controlled trial. JAMA 2002;288(3):321–33.

[33] LaCroix AZ. Estrogen with and without progestin: benefits and risks of short-term use. Am J Med 2005;118(12 Suppl. 2):79–87.

[34] Maricic M, Adachi JD, et al. Early effects of raloxifene on clinical vertebral fractures at 12 months in postmenopausal women with osteoporosis. Arch Intern Med 2002;162(10):1140–3.

[35] Black DM, Greenspan SL, et al. The effects of parathyroid hormone and alendronate alone or in combination in postmenopausal osteoporosis. N Engl J Med 2003;349(13):1207–15.

[36] Ring JD, Orwoll ES, et al. Treatment of male osteoporosis: recent advances with alendronate. Osteoporos Int 2002;13(3):195–9.

[37] Odvina CV, Zerwekh JE, Rae DS, et al. Severely suppressed bone turnover: a potential complication of alendronate therapy. J Clin Endocrinol Metab 2005;900(3):1897–9.

[38] Agarwala S, Jain D, Joshi VR, et al. Efficacy of alendronate, a bisphosphonate, in the treatment of AVEN of the hip: a prospective open-label study. Rheumatology (Oxford) 2005;44(3):352–9.

[39] Chestnut CH III, Silverman S, Andriano K, et al. A randomized trial of nasal spray salmon calcitonin in postmenopausal women with established osteoporosis: the prevent recurrence of osteoporotic fractures study. Am J Med 2000;109(4):267–76.

[40] Ofluoglu D, Akyuz G, Unay O, et al. The effect of calcitonin on beta-endorphin levels in postmenopausal osteoporotic patients with back pain. Clin Rheumatol 2006.

[41] Dempster DW, Cosman F, et al. Effects of daily treatment with parathyroid hormone on bone microarchitecture and turnover in patients with osteoporosis: a paired biopsy study. J Bone Miner Res 2001;16(10):1846–53.

[42] Neer RM, Arnaud CD, et al. Effect of parathyroid hormone (1–34) on fractures and bone mineral density in postmenopausal women with osteoporosis. N Engl J Med 2001;344(19):1434–41.

[43] Body JJ, Gaich GA, et al. A randomized double-blind trial to compare the efficacy of teriparatide [recombinant human parathyroid hormone (1–34)] with alendronate in postmenopausal women with osteoporosis. J Clin Endocrinol Metab 2002; 87(10):4528–35.

[44] Kurland ES, Heller SL, Diamond B, et al. The importance of bisphosphonate therapy in maintaining bone mass in men after therapy with teriparatide. Osteoporosis Int 2004;15(12):992–7.

[45] Finkelstein JS, Hayes A, et al. The effects of parathyroid hormone, alendronate, or both in men with osteoporosis. N Engl J Med 2002;349(13):1216–26.

[46] Cosman F, Nieves J, Zion M, et al. Daily and cyclic parathyroid hormone in women receiving alendronate. N Engl J Med 2005;353(6):566–75.

[47] Ste-Marie LG, Schwartz SL, Hossain A, et al. Effect of teriparatide on BMD when given to postmenopausal women receiving hormone replacement therapy. J Bone Miner Res 2006;21(2): 283–91.

[48] Deal C, Omizo M, Schwartz EN, et al. Combination teriparatide and raloxifene therapy for postmenopausal osteoporosis: results from a 6-month double-blind placebo-controlled trial. J Bone Miner Res 2005;20(11):1905–11.

[49] Alkhiary YM, Gerstenfeld LC, Krall E, et al. Enhancement of experimental fracture-healing by systemic administration of recombinant human parathyroid hormone. J Bone Joint Surg Am 2005; 87(4):731–41.

[50] Indications and reporting for dual-energy x-ray absorptiometry. J Clin Densitom 2004;7:37.

Hip Fracture and Its Consequences: Differences Between Men and Women

Denise L. Orwig, PhD*, Julia Chan, BA, Jay Magaziner, PhD, MSHyg

Division of Gerontology, Department of Epidemiology and Preventive Medicine, University of Maryland School of Medicine, 660 West Redwood Street, Suite 200, Baltimore, MD 21201, USA

Hip fractures are a recognized public health problem in women; however, they are also a significant problem in men, especially in light of the aging of the population. Approximately 25% to 30% of the 330,000 hip fractures in the United States occur in men [1,2], and in 2004, approximately 93,000 hip fractures were reported in men aged 65 and older [3]. By 2025, the annual incidence of hip fracture in men is anticipated to approximate that currently seen among women, and by 2050 it is conservatively projected that 2.5 million hip fractures will occur worldwide each year in men [1,4]. Relatively few men have been included in studies of osteoporosis and hip fracture, so little is known about fracture consequences in men. Clinical trials of osteoporosis treatments have been conducted primarily in women, so evidence-based therapeutic options for men are limited. Because of this lack of evidence about men and the fact that the most hip fractures occur in white women, guidelines for diagnosis and management of osteoporosis and its consequences are generally limited to postmenopausal white women [5–13]. We also cannot generalize from what is known about hip fracture in women and assume that this information is equally pertinent to the experiences of men who fracture [14]. The National Osteoporosis Foundation and the National Institutes of Health have stressed the relative deficits in our knowledge of osteoporosis and its consequences in men [6,8].

This article describes the state of knowledge regarding gender differences with respect to hip fracture and its subsequent outcomes. Most of the work to date investigating hip fracture patients has been done with women, yet some evidence from a few studies with a significant number of male hip fracture patients and from nonfracture samples suggests that women and men may be different at the time of fracture and will have a different course of recovery.

Hip fracture

A hip fracture is a sentinel event in the lives of older adults and causes severe consequences with respect to increased mortality, morbidity, functional dependence, incidence of other health conditions, and costs. As the proportion of the population aged 65 and older continues to expand, the absolute number of hip fractures is expected to increase significantly over the next several decades. Approximately one in four hip fractures occurs in men, yet there is a consistent age-specific increase in hip fracture incidence in men, whereas in women the rate seems to stabilize [15,16]. The lifetime incidence of hip fracture at age 50 is 17% to 22% in women and 6% to 11% in men [17,18]. The reasons for the lower incidence of hip fracture in men compared with women are related to (1) men having a higher peak bone density, (2) men losing less

---

Funding for this work was provided, in part, by The National Institute on Aging (NIA) Claude D. Pepper Older Americans Independence Center P60-AG12583 and Epidemiology of Aging Training Grant T32-AG00262.

* Corresponding author.

 E-mail address: dorwig@epi.umaryland.edu (D.L. Orwig).

bone during aging, (3) men not becoming hypogonadal, (4) men sustaining fewer falls, and (5) men having a shorter life span [19].

It is generally accepted that to sustain a hip fracture, one needs to have weak bones and a stress on bone of sufficient force to produce a break [20]. Most people with hip fracture have osteoporosis, which is defined as a condition characterized by low bone mass and microarchitectural deterioration of bone tissue, with consequent increase in bone fragility and fracture risk. Prevalence of osteoporosis in men over the age of 50 is 3% to 6% compared with 13% to 18% in women [21]. In 2002, more than 14 million men in the United States were estimated to have low bone mass and osteoporosis [22]. Compared with women who fracture a hip, men also tend to have a higher prevalence of risk factors for osteoporosis, including excessive alcohol consumption, current smoking, and low calcium intake [19,23].

Ninety-five percent of all hip fractures are attributed to falls [24]. Falls are the leading cause of nonfatal injury in the elderly population, with one in every three persons over the age of 65 falling each year [25–27]; one of the strongest predictors of falling is having a previous fall, which can increase the chance of another fall by as much as threefold [28]. Although women are more likely to fall and fracture in general, men seem to be more likely to fall after fracture [29]. More importantly, as many as 12% of people who suffer from a hip fracture are at risk of experiencing a subsequent fall [29]. Sex differences in falls also may be attributed to where the fall occurs. Although women are more likely to fall, these falls tend to be indoors, whereas men are more likely to fall outdoors. The difference in fall location also has been associated with differences in frailty; individuals who are frail tend to fall indoors and people who are more active and overexert themselves tend to fall outdoors. It may be functional limitations and environment factors versus sex itself that lead to varying patterns of falls between men and women.

Although men are less likely to have weak bones and are less likely to fall, it may be that men and women who are frailer may be more restricted to doing activities in their own homes, which is where they fall, and, in the presence of osteoporosis, are at greatest risk of sustaining a fracture.

Outcomes

Over the past 20 years, research on the consequences of hip fracture has progressed to understand the survival, functional, and physiologic sequelae of hip fracture. Because most hip fractures have occurred in white women, most of this work has focused on this group. The remainder of this article provides information on the differential consequences of hip fracture in men versus women, when that information is available.

Mortality

Research has documented that there is excess mortality after a hip fracture in women [30] and in men [31], with mortality after a hip fracture being higher in men compared with women for at least 2 years after fracture [32–41]. Mortality after a hip fracture in women is estimated to increase 12% to 20% [42], whereas the mortality experienced by men is twice as high [40,43–45]. Using data from a cohort of hip fracture patients, Wehren and colleagues [40] conducted a series of analyses to evaluate differences in causes of death between men and women over a 2-year period. Men were twice as likely as women to die during the first and second years after a hip fracture (OR 2.28; confidence interval:1.47–3.54 at 1 year and OR 2.21; confidence interval:1.48–3.31 at 2 years), although these differences were not explained by prefracture comorbidity, fracture type, surgical procedure, or operative complications (Table 1). This finding suggests that a new process is set in motion after a fracture that differs in men and women. Although elevations in causes of death were comparable for men and women for most major causes, the greatest increases in mortality for men versus women, relative to age- and sex-specific rates in the general population, were seen for infectious causes of death (septicemia and pneumonia).

This work demonstrates that survival after hip fracture is lower in men than in women, even after accounting for differences in health status at the time of fracture. Perhaps more remarkable is the discrepancy in the causes of death, which suggests that the route of entry for infectious pathogens or susceptibility to infection after the fracture differs for men and women. One hypothesis is that disease susceptibility increases in men more than women after a fracture; another is that the infectious agents enter the men's systems differentially from the way they enter women's.

Morbidity

The greater mortality seen in men who fracture has been linked to younger age at time of fracture

Table 1
Rate ratio of observed to expected deaths among men and women after hip fracture (95% CI)

Cause of death	Year 1		Year 2	
	Men	Women	Men	Women
Cardiovascular diseases	3.57 (2.02, 6.28)	3.82 (3.62, 4.03)	1.30 (0.42, 4.02)	2.03 (1.91, 2.15)
Cancer	3.65 (1.90, 7.02)	1.04 (0.95, 1.14)	0.59 (0.08, 4.19)	1.23 (1.13, 1.34)
Cerebrovascular disease	3.12 (0.78, 12.46)	2.53 (2.26, 2.82)	6.80 (2.19, 21.07)	2.13 (1.90, 2.38)
COPD	3.34 (0.83, 13.38)	3.30 (2.86, 3.81)	4.86 (1.12, 19.44)	[a]
Pneumonia, influenza	23.81 (12.81, 44.25)	7.46 (6.51, 8.55)	10.38 (3.35, 32.19)	4.00 (3.46, 4.62)
Diabetes mellitus	4.06 (0.58, 28.80)	1.14 (0.91, 1.43)	5.90 (0.83, 41.86)	[a]
Renal (including failure)	23.92 (7.71, 74.16)	[a]	[a]	9.36 (7.17, 12.22)
Septicemia	87.91 (16.49, 175.80)	36.72 (28.16, 47.87)	31.95 (7.99, 127.76)	13.33 (10.16, 17.48)
Overall mortality	5.19 (2.77, 9.74)	3.24 (2.18, 4.82)	1.31 (0.69, 2.47)	1.55 (1.01, 2.38)
Overall, excluding infections[b]	3.46 (1.79, 6.67)	2.47 (1.63, 3.72)	0.96 (0.48, 1.91)	1.26 (0.80, 1.98)

Abbreviation: COPD, Chronic obstructive pulmonary disease.
[a] No deaths from that cause during time period.
[b] Infections include pneumonia, influenza, and septicemia.

[44,46] and to a greater comorbidity burden at the time of fracture [39,44,47–49]. Poor and colleagues [39] reported that the excess mortality among men was better explained by the interaction of the fracture with the underlying comorbidity status because survival was worse among male hip fracture patients compared with age-matched controls given any level of comorbidity [39].

A recent study highlighted the comorbidity burden in men who fracture versus women. Of 674 recruited hip fracture patients, 522 were women and 152 were men, of whom 350 women and 82 men were evaluated for follow-up function at 1 year. Table 2 shows some sex differences at the time of fracture. The average age for women in this study was 81.7 years (SD = 7.4) and for men it was 79.1 years (SD = 7.4). The overall summary score for comorbidity showed a greater comorbidity burden for the men in this sample, with an average score of 1.65 (SD = 1.45) for women and 2.57 (SD = 1.98) for men. Women at admission had higher rates of diagnosed arthritis (30.3% versus 20.4% for men), hypertension (48.5% versus 34.9%), and osteoporosis (13.6% versus 4.6%). Men had higher rates of chronic obstructive pulmonary disease, bronchitis, emphysema, or asthma (26.3% versus 15.3%), myocardial infarction (9.8% versus 18.4%), and stroke (20.4% versus 7.9%). Men were more dependent in walking 10 feet (36.3% versus 28.7%)

Table 2
Descriptive statistics of baseline characteristics of hip fracture patients by sex

Characteristics	Men (n = 152)	Women (n = 522)
Age X̄ (SD)	79.1 years (7.39)	81.7 years (7.37)
% Non-white	7.9	7.3
Pre-fx lower extremity ADL[a] X̄ (SD)	3.57 (3.34)	3.26 (3.41)
Pre-fx instrumental ADL[a] X̄ (SD)	3.17 (2.66)	2.96 (2.33)
Pre-fx walk 10 feet (% dependent)	36.3	28.7
Pre-fx walk 1 block (% dependent)	43.9	41.9
In-hospital cognitive status (MMSE) X̄ (SD)	27.42 (6.26)	27.25 (6.22)
In-hospital depression (CES-D) X̄ (SD)	18.57 (11.14)	16.32 (10.34)
Comorbidity[b] X̄ (SD)	2.57 (1.98)	1.65 (1.45)
% Subcapital fracture	47.3	45.1

Abbreviations: ADL, activities of daily living; CES-D, Center for Epidemiological Studies Depression Scale; MMSE, Mini-Mental State Exam [151].
[a] Number of dependencies reported.
[b] Scale adapted from Charlson and colleagues [152].

and had slightly more symptoms of depression during their hospital stay (Center for Epidemiological Studies Depression Scale [CES-D] score: 18.6 versus 16.3).

These data suggest that men and women who fracture have a different health status at the time of fracture. Men are younger and sicker, and the differences in rates of specific comorbidities may have implications for types of treatment required and ability to tolerate certain interventions.

Subsequent fractures

The contribution of hip fracture to the risk of future fracture was assessed using a hip fracture cohort and data from all Established Populations for the Epidemiological Studies of the Elderly sites [50]. Self-reported non–hip fracture was the primary outcome. After adjusting for race, sex, body mass index, cognitive impairment, and selected comorbidities and functional impairments, hip fracture patients had 1.62-fold increased risk of subsequent fractures. The risk of suffering a second hip fracture also is greater in men than in women [51,52].

Physical function

After a hip fracture, there is a sudden loss of physical function (as measured by ability to walk independently and walking speed) followed by a period of recovery that can continue for a year or more, depending on the area of function examined [53,54]. As much as 80% of individuals who survive a hip fracture fail to regain their functional independence, and almost 50% of men with hip fractures must be institutionalized because of the fracture [39]. By the end of a year, half of the patients who were able to walk independently before their fracture were not able to walk independently [53,54], which translates into an excess loss of function of approximately 25% when compared with loss in similar persons who do not have a hip fracture [55]. Additional data also demonstrate that in women, walking speed is approximately one standard deviation lower than that of a similar group of women who do not sustain a hip fracture [56]. Sex differences have not been reported for postfracture function, although the few studies that did consider sex differences may have been too small to identify differences even if they did exist [54,57–60].

Hawkes and colleagues [60] conducted longitudinal analyses of the course of recovery among hip fracture patients and showed some sex differences. For example, 59.8% of men were impaired in walking 10 feet by self-report at 12 months compared with 52% of women, whereas the average number of lower extremity physical activities of daily living impaired at 12 months was 7.1 for men versus 6.4 for women. Fig. 1 shows trends in sex differences for six areas of physical and instrumental function that require lower extremities. Negative effects for men relative to women are small, although they do suggest that men fare more poorly, which is shown on the y-axis. These differences are largely gone by 24 months.

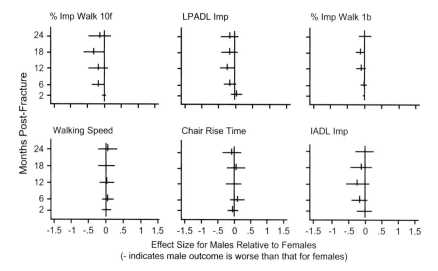

Fig. 1. Adjusted past-fracture differences according to sex.

Physical function declines over time in older men and women [61–65], and as with loss of strength, women exhibit more rapid declines than men [62–65]. Gender does not seem associated with disability; however, women may be more likely to experience moderate disability and men to experience severe or mild disability [66,67]. It also has been reported that women are less likely to recover function after disability occurs [64].

Psychosocial factors

Most prior research on the relationship between psychosocial factors and recovery after hip fracture has been done in women. When men were included in these studies the numbers were not sufficiently large to permit evaluation of sex differences [68–70]. It is essential, however, to explore the trajectory of cognition, mood, and social function after hip fracture in men. Affective and cognitive status change dramatically after a fracture, each reaching a prevalence as high as 50% immediately after the event [53]. Incidence of depression is higher for men and women who have had a hip fracture [43,71,72], and it has repeatedly been related to poor recovery [68,73–75], as have post-fracture and 2-month cognition [76,77]. Social functioning is important and changes after a hip fracture [53]; it may influence other functional outcomes and participation in rehabilitation and physical activity [54,70,78–82]. Because many of these psychosocial factors may operate differently in men and women who have fractured, it is important to identify these differences in the hip fracture population so that the most appropriate strategies to foster recovery can be developed and implemented.

Bone mineral density and bone strength

There is a strong association between bone mineral density (BMD) measured at the femoral neck or total hip and hip fracture risk. Nearly all women in a series of cohorts in the Baltimore Hip Studies had low hip BMD (classifiable as osteoporosis by WHO criteria) at the time of fracture [13]. Profound losses in BMD (5%–8% loss at the contralateral femoral neck) during the year after fracture have been reported [83–85]. Although two of these studies included small numbers of men, sex-specific data were not presented [83,84]. Some reports also have shown that men fracture at the same absolute level of BMD as women [86], whereas others have not found this to be the case [87].

Despite the strong association between BMD and hip fracture risk, several other characteristics of bone also relate to hip fracture risk. One such example is that the size and shape of bones (ie, bone geometry) have effects on bone strength and fracture risk that may not be apparent in BMD [88–96]. Men have, on average, larger bones and higher BMD than women and different hormonal and metabolic environments [97]. Progressive loss of bone strength with advancing age has been reported in men and women [88,89,98,99]. Compensatory geometric adaptations in both sexes may mitigate bone loss, and it is conceivable that decrements in strength may be greater in women [90,92,98,99].

Hip structural analysis, a technique for estimating geometry from bone mineral scans of the hip, has shown that decline in BMD with age occurs by bone loss and periosteal apposition [100]. The latter effect reduces BMD by broadening the outer margins of the region over which the mass is averaged in calculating the density measurement. Expansion of bone diameter seems to preserve section modulus, an index of bending strength, despite loss of bone mass. Although BMD declines more rapidly with age in older women, the difference may be caused by faster expansion of diameter, not more rapid bone loss [101]. Engineers who design light-weight tubular structures recognize that a tubular structure can maintain its bending resistance with less material by expanding the outer diameter, but care must be taken to ensure that the wall is not thinned to the point at which it might buckle (crumple) under compressive loads. Strength begins to be compromised when the ratio of outer radius/wall thickness (buckling ratio) exceeds a factor of ten [102].

Mother Nature seems to be using the same lightening principle in preserving bending resistance in aging bones but may not be as careful in avoiding geometries that might lose strength caused by buckling. Bone mineral scans do not provide enough information to reliably measure cortical thickness, especially in thin cortex regions, but a crude estimate of buckling ratio is generated by the hip structural analysis method using the measured outer diameter and simple annulus or cylindrical shape models of the cortex. Estimated femoral neck buckling ratios generally increase with age through adulthood [99], are higher in men and women who suffer hip fractures [103], and are reduced by osteoporosis treatments [104–106]. Structural parameters of bone in men at the time of fracture and subsequent adaptation of bone and body composition differ from those observed in women [107–109]. Analysis of data from the contralateral hip in women a year after hip fracture has shown important degradation in strength and increased instability of the proximal femur [110]. Detailed studies of bone strength and structure after hip fracture in men are not currently available.

Muscle and fat mass

Substantial loss of muscle mass, 5% to 6%, occurs in women during the year after hip fracture, and fat mass increases by 4% to 11% over the same time period [83,85]. Small numbers of men were included in one of these studies [83], but because sex-specific data were not presented, the effects of hip fracture on soft tissue compartments in men are undetermined currently. Average grip strength measured in women during hospitalization after hip fracture is relatively low but improves by 2 months later. Accelerated losses of muscle mass and strength after hip fracture may contribute to future fracture risk and slower recovery of function. Progressive loss of muscle mass with aging has been associated with decreasing physiologic and functional reserve [111], and accelerated or greater losses may indicate the development of sarcopenia [112]. Changes in lean and fat mass of persons aged 70 years and older have been reported to be greater in men than in women [113], but whether the same would occur after a hip fracture has not yet been evaluated.

Muscle strength has been reported to be more strongly associated with mechanical properties of bone than muscle mass in men and women [114]. In addition, strength is a major component of balance and gait, so losses of strength may have important effects on risk of falls and fracture [115]. This is especially critical for hip fracture patients, in whom gait and balance are affected by the fracture. Grip strength correlates well with overall muscle strength [116,117] and predicts later disability in men [118] and short-term mortality risk in elderly women [119]. Grip strength is greater in men than in women but decreases progressively with advancing age in both sexes [120]. The rate of decline, however, is greater in women [121]. Data also have demonstrated that recovery of gait and balance precedes recovery of physical function [53]. Muscle quality deteriorates more in women than men with aging [122], and targeted exercise has been reported to have different effects in men and women [123], which may translate into differences in functioning.

Post-fracture management

Rehabilitation

Because fractures are associated with global declines in physical performance and function, even among high-functioning men and women [124], physical rehabilitation is an important focus of post-fracture care. Participation in rehabilitation activities and ongoing physical activity can have an important impact on the recovery process after a hip fracture [125–130]. A recent study compared rehabilitation outcomes among male and female hip fracture patients and found that gender did not affect the success of rehabilitation at the end of hospitalization [131]. Another component of rehabilitation is physical activity. Although prior studies include mostly women and do not consider sex differences, there is some evidence of sex differences in physical activity in older adults [65]. The prevalence of inactivity among white women aged 75 and older is 47% and among older white men is 37% [132].

Pharmacologic treatment

The National Osteoporosis Foundation treatment guidelines for osteoporosis recommend pharmacologic intervention (including calcium and vitamin D, raloxifene, calcitonin, alendronate, risendronate) in female patients who have a history of either vertebral or hip fracture, regardless of BMD [133,134], and no recommendations for treatment in men. There are only two US Food and Drug Administration–approved therapies for treatment of osteoporosis in men: teriparatide [135] and alendronate [136]. No pharmacologic osteoporosis treatments have been tested in hip fracture patients, however, so effectiveness only can be inferred from results in other populations.

Despite recommendations, most hip fracture patients do not receive definitive pharmacologic treatment, nor is osteoporosis evaluation generally performed [137–144]. Although undertreatment is recognized in women, it would be expected to be more common in men, with few receiving treatment [52]. Osteoporosis diagnosis, which increases the likelihood of treatment [138, 145–147], is made in less than 20% of women who sustain a hip fracture, even after the event [138,140,144,147]. General treatment rates less than 20% are typical, even as long as 1 year after the fracture [138–144,148,149], and less aggressive calcium with vitamin D supplementation is the most commonly used treatment [137,149]. One particular study investigated treatment of osteoporosis after a hip fracture in men and found that only 1% of men were discharged with treatment other than calcium and vitamin D compared with 25% of women [150]. Over the 5-year follow-up period, only 27% of the men had commenced treatment for osteoporosis (including calcium and vitamin D) versus 71% of the women, and most of these patients had never had a BMD test.

Summary

Overall, although approximately one fourth of all hip fractures occur in men, it seems that there are more unanswered questions than definitive information about the sequelae of hip fractures in this group of hip fracture patients. Still, it seems clear that men are generally sicker than women when they fracture a hip and that their chances for survival are half of what they are in women. Evidence also indicates that men die differentially of infection, whereas men and women both die at a somewhat higher rate of the other causes common to older persons. Evidence for a pattern of functional recovery in men versus women is equivocal, and evidence for post-fracture physiologic differences between men and women does not exist.

To fracture a hip, two conditions must be met: (1) one must have weak bones, as in osteoporosis and (2) these weak bones must be stressed to a breaking point, as occurs with a fall. Not surprisingly, osteoporosis and falls before a first fracture are more common in women than in men, which likely accounts for much of the gender differential in fracture rates. Why are men who fracture sicker than women, and why do they die at a higher rate? Although speculative, it is likely that to develop weak bones, other disease processes must be operating in men than in women, who lose bone naturally throughout their adult life and at an accelerated rate after menopause. These disease processes appear as increased levels of comorbidity and functional deficits at the time of fracture in men. It may be that these underlying disease processes also promote falls in vulnerable individuals, thereby leading to fractures and the poorer survival rate for men. Increased rates of infection may be caused by compromised immunity as a result of these underlying problems or an environmental source. The equivocal findings on sex differences in functional outcomes may be caused by the fact that studies to date have been too small to detect these differences, or they may be real. For example, because studies of post-fracture functioning are restricted to persons who survive, it is possible that the men and women who survive the first few months after a fracture are really similar and that their recovery occurs in a similar manner.

Given the differences we do see, it is clear that men and women who sustain fractures are different and have different needs and outcomes, at least in the first months after the fracture. This in itself has important implications for individuals who manage the care of these patients. The limited evidence on these differences and their implications for care also signals a need to obtain more information to improve the care of men and women who fracture a hip and recognize that at least some of what we know about the management of fractures in women may not apply directly to men without modifications that take sex differences in status at the time of fracture and likelihood of survival into account.

Acknowledgements

The authors would like to thank Thomas Beck, PhD, for his thoughtful review of portions of this article.

References

[1] Eastell R, Boyle IT, Compston J, et al. Management of male osteoporosis: report of the UK Consensus Group. Q J Med 1998;91:71–92.

[2] Orwoll ES, Klein RF. Osteoporosis in men. Endocr Rev 1995;16(1):87–116.

[3] DeFrances CJ, Podgornik MN. 2004 national hospital discharge survey: annual summary with detailed diagnosis and procedure data. Advance data from vital and health statistics; 2006. Available at: http://www.cdc.gov/nchs/data/ad/ad371.pdf.

[4] Seeman E. The dilemma of osteoporosis in men. Am J Med 1995;98(Suppl 2A):76S–88S.

[5] Brown JP, Josse RG. 2002 clinical practice guidelines for the diagnosis and management of osteoporosis in Canada. Can Med Assoc J 2002;167(10 Suppl):S1–34.

[6] National Osteoporosis Foundation. Osteoporosis: review of the evidence for prevention, diagnosis, and treatment and cost-effectiveness analysis. Osteoporos Int 1998;8(Suppl 4):S1–88.

[7] Nelson HD, Helfand M, Woolf SH, et al. Screening for postmenopausal osteoporosis: a review of the evidence for the US Preventive Services Task Force. Ann Intern Med 2002;137:529–41.

[8] NIH Consensus Development Panel. Osteoporosis prevention, diagnosis, and therapy. JAMA 2001; 285:785–95.

[9] Sambrook PN, Seeman E, Phillips SR, et al. Preventing osteoporosis: outcomes of the Australian Fracture Prevention Summit. Med J Aust 2002; 176(Suppl):S1–16.

[10] US Preventive Services Task Force. Screening for osteoporosis in postmenopausal women: recommendations and rationale. Ann Intern Med 2002; 137:526–8.

[11] Working Party Writing Group. Summary and recommendations of the report "Osteoporosis: clinical guidelines for the prevention and treatment." London: Royal College of Physicians; 1999.

[12] Working Party Writing Group. Osteoporosis: clinical guidelines for prevention and treatment. Update on pharmacological interventions and an algorithm for management. London: Royal College of Physicians; 2001.

[13] World Health Organization. Assessment of fracture risk and application to screening for postmenopausal osteoporosis. Geneva, Switzerland: WHO Technical Report Series; 1994.

[14] Institute of Medicine. Exploring the biological contributions to human health: does sex matter? Washington, DC: National Academy Press; 2001.

[15] Chang KP, Center JR, Nguyen TV, et al. Incidence of hip and other osteoporotic fractures in elderly men and women: Dubbo Osteoporosis Epidemiology Study. J Bone Miner Res 2004; 19(4):532–6.

[16] Haentjens P, Johnell O, Kanis JA, et al. Evidence from data searches and life-table analyses for gender-related differences in absolute risk of hip fracture after Colles' or spine fracture: Colles' fracture as an early and sensitive marker of skeletal fragility in white men. J Bone Miner Res 2004; 19(12):1933–44.

[17] Melton LJ III, Atkinson EJ, O'Connor MK, et al. Bone density and fracture risk in men. J Bone Miner Res 1998;13(12):1915–23.

[18] Oden A, Dawson A, Dere W, et al. Lifetime risk of hip fractures is underestimated. Osteoporos Int 1998;8:599–603.

[19] Seeman E. Osteoporosis in men: epidemiology, pathophysiology, and treatment possibilities. Am J Med 1993;95(5A):22S–8S.

[20] Mayhew PM, Thomas CD, Clement JG, et al. Relation between age, femoral neck cortical stability, and hip fracture risk. Lancet 2005;366(9480): 129–35.

[21] Looker AC, Orwoll ES, Johnston CC, et al. Prevalence of low femoral bone density in older US adults from NHANES III. J Bone Miner Res 1997;12(11):1761–8.

[22] National Osteoporosis Foundation. NOF prevalence report. America's bone health: the state of osteoporosis and low bone mass in our nation. Osteoporosis and low bone mass: males. Available at: http://www.nof.org/advocacy/prevalence/index.htm. Accessed July 20, 2006.

[23] Diamond TH, Thornley SW, Sekel R, et al. Hip fracture in elderly men: prognostic factors and outcomes. Med J Aust 1997;167:412–5.

[24] Nyberg L, Gustafson Y, Berggren D, et al. Falls leading to femoral neck fractures in lucid older people. J Am Geriatr Soc 1996;44:156–60.

[25] Duthie EH Jr. Falls. Med Clin North Am 1989; 73(6):1321–36.

[26] Tinetti ME, Speechley M, Ginter SF. Risk factors for falls among elderly persons living in the community. N Engl J Med 1988;319(26):1701–7.

[27] Sattin RW. Falls among older persons: a public health perspective. Annu Rev Public Health 1992; 13:489–508.

[28] American Geriatrics Society, British Geriatrics Society, and American Academy of Orthopaedic Surgeons Panel on Falls Prevention. Guideline for the prevention of falls in older persons. J Am Geriatr Soc 2001;49(5):664–72.

[29] Pils K, Neumann F, Meisner W, et al. Predictors of falls in elderly people during rehabilitation after hip fracture: who is at risk of a second one? Z Gerontol Geriatr 2003;36(1):16–22.

[30] Magaziner J, Lydick E, Hawkes W, et al. Excess mortality attributable to hip fracture in white women aged 70 years and older. Am J Public Health 1997;87:1630–6.

[31] Fransen M, Woodward M, Norton R, et al. Excess mortality or institutionalization after hip fracture: men are at greater risk than women. J Am Geriatr Soc 2002;50(4):685–90.

[32] Cooper C, Campion G, Melton LJ. Hip fractures in the elderly: a world-wide projection. Osteoporos Int 1992;2:285–9.

[33] Mullen JO, Mullen NL. Hip fracture mortality: a prospective, multifactorial study to predict and minimize death risk. Clin Orthop 1992;280:214–22.

[34] Huuskonen J, Kroger H, Arnala I, et al. Characteristics of male hip fracture patients. Ann Chir Gynaecol 1999;88:48–53.

[35] Siris ES, Miller PD, Barrett-Connor E, et al. Identification and fracture outcomes of undiagnosed low bone mineral density in postmenopausal women: results from the National Osteoporosis Risk Assessment. JAMA 2001;286:2815–22.

[36] Trivedi DP, Khaw KT. Bone mineral density at the hip predicts mortality in elderly men. Osteoporos Int 2001;12:259–65.

[37] Genant HK, Cooper C, Poor G, et al. Interim report and recommendations of the World Health Organization task-force for osteoporosis. Osteoporos Int 1999;10:259–64.

[38] Black DM, Cummings SR, Karpf DB, et al. Randomised trial of effect of alendronate on risk of fracture in women with existing vertebral fractures. Lancet 1996;348(9041):1535–41.

[39] Poor G, Atkinson EJ, O'Fallon WM, et al. Determinants of reduced survival following hip fractures in men. Clin Orthop Relat Res 1995;319:260–5.

[40] Wehren LE, Hawkes W, Orwig D, et al. Gender differences in mortality after hip fracture: the role of infection. J Bone Miner Res 2003;18(12):2231–7.

[41] Trombetti A, Herrmann F, Schurch MA, et al. Survival and potential years of life lost after hip fracture in men and age-matched women. Osteoporos Int 2002;13:731–7.

[42] Cummings SR, Kelsey JL, Nevitt MC, et al. Epidemiology of osteoporosis and osteoporotic fractures. Epidemiol Rev 1985;7:178–208.

[43] Magaziner J, Simonsick EM, Kashner TM, et al. Survival experience of aged hip fracture patients. Am J Public Health 1989;79:274–8.

[44] Jacobsen SJ, Goldberg J, Miles TP, et al. Race and sex differences in mortality following fracture of the hip. Am J Public Health 1992;82:1147–50.

[45] Holt EM, Evans RA, Hindley CJ, et al. 1000 femoral neck fractures: the effect of pre-injury mobility and surgical experience on outcome. Injury 1994;25:91–5.

[46] Lofman O, Berglund K, Larsson L, et al. Changes in hip fracture epidemiology: redistribution between ages, genders and fracture types. Osteoporos Int 2002;13:18–25.

[47] Forsen L, Sogaard AJ, Meyer HE, et al. Survival after hip fracture: short- and long-term excess mortality according to age and gender. Osteoporos Int 1999;10:73–8.

[48] Marottoli RA, Berkman LF, Leo-Summers L, et al. Predictors of mortality and institutionalization after hip fracture: the New Haven EPESE cohort. Am J Public Health 1994;84(11):1807–12.

[49] Center JR, Nguyen TV, Schneider D, et al. Mortality after all major types of osteoporotic fracture in men and women: an observational study. Lancet 1999;353:878–82.

[50] Colón-Emeric C, Kuchibhatla M, Pieper C, et al. The contribution of hip fracture to risk of secondary fractures: data from two longitudinal studies. Osteoporos Int 2003;14(11):879–83.

[51] Melton LJI, Ilstrup DM, Beckenbaugh RD, et al. Hip fracture recurrence: a population-based study. Clin Orthop Relat Res 1982;167:131–8.

[52] Colon-Emeric CS, Sloane R, Hawkes WG, et al. The risk of subsequent fractures in community-dwelling men and male veterans with hip fracture. Am J Med 2000;109:324–6.

[53] Magaziner J, Hawkes W, Hebel JR, et al. Recovery from hip fracture in eight areas of function. J Gerontol A Biol Sci Med Sci 2000;55A:M498–507.

[54] Magaziner J, Simonsick EM, Kashner TM, et al. Predictors of functional recovery one year following hospital discharge for hip fracture: a prospective study. J Gerontol 1990;45(3):M101–7.

[55] Magaziner J, Fredman L, Hawkes W, et al. Changes in functional status attributable to hip fracture: a comparison of hip fracture patients to community-dwelling aged. Am J Epidemiol 2003;157:1023–31.

[56] Fredman L, Magaziner J, Hawkes W, et al. Female hip fracture patients had poorer performance-based functioning than community-dwelling peers over 2-year follow-up period. J Clin Epidemiol 2005;58(12):1289–98.

[57] Beloosesky Y, Grinblat J, Epelboym B, et al. Functional gain of hip fracture patients in different cognitive and functional groups. Clin Rehabil 2002;16(3):321–8.

[58] Koval KJ, Skovron ML, Aharonoff GB, et al. Ambulatory ability after hip fracture: a prospective study in geriatric patients. Clin Orthop 1995;(310):150–9.

[59] Hannan EL, Magaziner J, Wang JJ, et al. Mortality and locomotion 6 months after hospitalization for hip fracture: risk factors and risk-adjusted hospital outcomes. JAMA 2001;285:2736–42.

[60] Hawkes WG, Wehren L, Orwig D, et al. Gender differences in functioning after hip fracture. J Gerontol A Biol Sci Med Sci 2006;61A(5):495–9.

[61] Seeman TE, Charpentier PA, Berkman LF, et al. Predicting changes in physical performance in a high-functioning elderly cohort: MacArthur studies of successful aging. J Gerontol 1994;49(3):M97–108.

[62] Schroll M, Bjornsbo-Scroll K, Ferrt N, et al. Health and physical performance of elderly Europeans: SENECA Investigators. Eur J Clin Nutr 1996;50(Suppl 2):S105–11.

[63] Schroll M, Avlund K, Davidsen M. Predictors of five-year functional ability in a longitudinal survey of men and women aged 75 to 80: the 1914 population in Glostrup, Denmark. Aging (Milano) 1997;9(1–2):143–52.

[64] Beckett LA, Brock DB, Lemke JH, et al. Analysis of change in self-reported physical function among older persons in four population studies. Am J Epidemiol 1996;143(8):766–78.

[65] Bennett KM. Gender and longitudinal changes in physical activities in later life. Age Ageing 1998;27(Suppl 3):24–8.

[66] Bond J, Gregson B, Smith M, et al. Outcomes following acute hospital care for stroke or hip fracture: how useful is an assessment of anxiety or depression for older people? Int J Geriatr Psychiatry 1998;13(9):601–10.

[67] Cobey JC, Cobey JH, Conant L, et al. Indicators of recovery from fractures of the hip. Clin Orthop 1976;117:258–62.

[68] Kempen G, Sanderman R, Scaf-Klomp W, et al. The role of depressive symptoms in recovery from injuries to the extremities in older persons: a prospective study. Int J Geriatr Psychiatry 2003;18:14–22.

[69] Chen C, Heinemann A, Granger C, et al. Functional gains and therapy intensity during subacute rehabilitation: a study of 20 facilities. Arch Phys Med Rehabil 2002;83:1514–23.

[70] Ottenbacher K, Smith P, Illig S, et al. Disparity in health services and outcomes for persons with hip

[70] fracture and lower extremity joint replacement. Med Care 2003;41:232–41.
[71] Mutran EJ, Reitzes DC, Mossey J, et al. Social support, depression, and recovery of walking ability following hip fracture surgery. J Gerontol Soc Sci 1995;50B(6):S354–61.
[72] Mossey JM, Knott K, Craik R. The effects of persistent depressive symptoms on hip fracture recovery. J Gerontol 1990;45(5):M163–8.
[73] Mossey JM, Mutran E, Knott K, et al. Determinants of recovery 12 months after hip fracture: the importance of psychosocial factors. Am J Public Health 1989;79(3):279–86.
[74] Zimmerman SI, Smith HD, Gruber-Baldini A, et al. Short-term persistent depression following hip fracture: a risk factor and target to increase resilience in elderly people. Soc Work Res 1999;23:187–96.
[75] Scaf-Klomp W, Sanderman R, Ormel J, et al. Depression in older people after fall-related injuries: a prospective study. Age Ageing 2003;32(1):88–94.
[76] Dolan MM, Hawkes WG, Zimmerman SI, et al. Delirium on hospital admission in aged hip fracture patients: prediction of mortality and 2-year functional outcomes. J Gerontol 2000;55A:M527–34.
[77] Gruber-Baldini A, Zimmerman S, Morrison R, et al. Cognitive impairment in hip fracture patients: timing of detection and longitudinal follow-up. J Am Geriatr Soc 2003;51(9):1227–36.
[78] Resnick B, Daly MP. Predictors of functional ability in geriatric rehabilitation patients. Rehabil Nurs 1998;23(1):21–9.
[79] Resnick B, Orwig D, Magaziner J, et al. The effect of social support on exercise behavior in older adults. Clin Nurs Res 2002;11(1):52–70.
[80] Chogahara M. A multidimensional scale for assessing positive and negative social influences on physical activity in older adults. J Gerontol B Psychol Sci Soc Sci 1999;54(6):S356–67.
[81] King AC, Rejeski WJ, Buchner DM. Physical activity interventions targeting older adults: a critical review and recommendations. Am J Prev Med 1998;15(4):316–33.
[82] Courneya KS, McAuley E. Cognitive mediators of the social influence-exercise adherence relationship: a test of the theory of planned behavior. J Behav Med 1995;18(5):499–515.
[83] Karlsson M, Nilsson JA, Sernbo I, et al. Changes of bone mineral mass and soft tissue composition after hip fracture. Bone 1996;18(1):19–22.
[84] Dirschl DR, Henderson RC, Oakley WC. Accelerated bone mineral loss following a hip fracture: a prospective longitudinal study. Bone 1997;21(2):79–82.
[85] Fox KM, Magaziner J, Hawkes WG, et al. Loss of bone density and lean body mass after hip fracture. Osteoporos Int 2000;11(1):31–5.
[86] de Laet CE, van Hout BA, Burger H, et al. Bone density and risk of hip fracture in men and women: cross sectional analysis. BMJ 1997;315(7102):221–5.
[87] Melton LJ III, Orwoll ES, Wasnich RD. Does bone density predict fractures comparably in men and women? Osteoporos Int 2001;12:707–9.
[88] Mosekilde L. Sex differences in age-related loss of vertebral trabecular bone mass and structure: biomechanical consequences. Bone 1989;10:425–32.
[89] Mosekilde L, Mosekilde L. Sex differences in age-related changes in vertebral body size, density, and biomechanical competence in normal individuals. Bone 1990;11:67–73.
[90] Cody DD, Divine GW, Nahigian K, et al. Bone density distribution and gender dominate femoral neck fracture risk predictors. Skeletal Radiol 2000;29:151–61.
[91] Cordey J, Schneider M, Belendez C, et al. Effect of bone size, not density, on the stiffness of the proximal part of normal and osteoporotic human femora. J Bone Miner Res 1992;7(Suppl 2):S437–44.
[92] Crabtree N, Lunt M, Holt G, et al. Hip geometry, bone mineral distribution, and bone strength in European men and women: the EPOS study. Bone 2000;27:151–9.
[93] Faulkner KG, Cummings SR, Black D, et al. Simple measurement of femoral geometry predicts hip fracture: the study of osteoporotic fractures. J Bone Miner Res 1993;8(10):1211–7.
[94] Martin B. Aging and strength of bone as a structural material. Calcif Tissue Int 1993;53(Suppl 1):S34–40.
[95] McCreadie BR, Golstein SA. Biomechanics of fracture: is bone mineral density sufficient to assess risk? J Bone Miner Res 2000;15:2305–8.
[96] Ferretti JL. Biomechanical properties of bone. In: Genant HK, Guglielmi G, Jergas M, editors. Bone densitometry and osteoporosis. Berlin: Springer-Verlag; 1998. p. 143–61.
[97] Looker AC, Beck TJ, Orwoll ES. Does body size account for gender differences in femur bone density and geometry? J Bone Miner Res 2001;16:1291–9.
[98] Beck TJ, Ruff CB, Scott WWJ, et al. Sex differences in geometry of the femoral neck with aging: a structural analysis of bone mineral data. Calcif Tissue Int 1992;50:24–9.
[99] Beck TJ, Looker AC, Ruff CB, et al. Structural trends in the aging femoral neck and proximal shaft: analysis of the third National Health and Nutrition Examination Survey dual-energy X-ray absorptiometry data. J Bone Miner Res 2000;15:2297–304.
[100] Beck TJ, Ruff CB, Warden KE, et al. Predicting femoral neck strength from bone mineral data: a structural approach. Invest Radiol 1990;25:6–18.
[101] Kaptoge S, Dalzell N, Loveridge N, et al. Effects of gender, anthropometric variables, and aging on the evolution of hip strength in men and women aged over 65. Bone 2003;32(5):561–70.

[102] Young W. Elastic stability formulas for stress and strain. In: Crawford TS, editor. Roark's formulas for stress and strain. New York: McGraw-Hill; 1989. p. 688.

[103] Duan Y, Beck TJ, Wang XF, et al. Structural and biomechanical basis of sexual dimorphism in femoral neck fragility has its origins in growth and aging. J Bone Miner Res 2003;18(10):1766–74.

[104] Greenspan SL, Beck TJ, Resnick NM, et al. Effect of hormone replacement, alendronate, or combination therapy on hip structural geometry: a 3-year, double-blind, placebo-controlled clinical trial. J Bone Miner Res 2005;20(9):1525–32.

[105] Uusi-Rasi K, Semanick LM, Zanchetta JR, et al. Effects of teriparatide [rhPTH (1–34)] treatment on structural geometry of the proximal femur in elderly osteoporotic women. Bone 2005;36(6): 948–58.

[106] Uusi-Rasi K, Beck TJ, Semanick LM, et al. Structural effects of raloxifene on the proximal femur: results from the multiple outcomes of raloxifene evaluation trial. Osteoporos Int 2006;17(4): 575–86.

[107] Seeman E. The structural basis of bone fragility in men. Bone 1999;25(1):143–7.

[108] Seeman E, Duan Y, Fong C, et al. Fracture site-specific deficits in bone size and volumetric density in men with spine or hip fractures. J Bone Miner Res 2001;16:120–7.

[109] Seeman E. During aging, men lose less bone than women because they gain more periosteal bone, not because they resorb less endosteal bone. Calcif Tissue Int 2001;69:205–8.

[110] Wehren LE, Beck TJ, Oreskovic TL, et al. Hip structural analysis in women with cervical and intertrochanteric fractures [abstract]. Osteoporos Int 2002;13(Suppl 1):S67.

[111] Fried LP, Tangen CM, Walston J, et al. Frailty in older adults: evidence for a phenotype. J Gerontol A Biol Sci Med Sci 2001;56A:M146–56.

[112] Evans WJ. What is sarcopenia? J Gerontol 1995; 50A(Special issue):5–8.

[113] Karlsson MK, Obrant KJ, Nilsson BE, et al. Changes in bone mineral, lean body mass and fat content as measured by dual energy X-ray absorptiometry: a longitudinal study. Calcif Tissue Int 2000;66:97–9.

[114] Hasegawa Y, Schneider P, Reiners C. Age, sex, and grip strength determine architectural bone parameters assessed by peripheral quantitative computed tomography (pQCT) at the human radius. J Biomech 2001;34:497–503.

[115] Wolfson L, Judge J, Whipple R, et al. Strength is a major factor in balance, gait, and the occurrence of falls. J Gerontol 1995;50A:64–7.

[116] Geusens P, Vandvyver C, Vanhoof J, et al. Quadriceps and grip strength are related to vitamin D receptor genotype in elderly nonobese women. J Bone Miner Res 1997;12:2082–8.

[117] Rantanen T, Era P, Kauppinen M, et al. Maximal isometric muscle strength and socioeconomic status, health, and physical activity in 75-year-old persons. J Aging Phys Act 1994;2:206–20.

[118] Rantanen T, Guralnik JM, Foley D, et al. Midlife hand grip strength as a predictor of old age disability. JAMA 1999;281(6):558–60.

[119] Phillips P. Grip strength, mental performance and nutritional status as indicators of mortality risk among female geriatric patients. Age Ageing 1986;15:53–6.

[120] Desrosiers J, Bravo G, Hebert R, et al. Normative data for grip strength of elderly men and women. Am J Occup Ther 1995;49:637–44.

[121] Bassey EJ. Longitudinal changes in selected physical capabilities: muscle strength, flexibility and body size. Age Ageing 1998;27(Suppl 3):12–6.

[122] Doherty TJ. The influence of aging and sex on skeletal muscle mass and strength. Curr Opin Clin Nutr Metab Care 2001;4(6):503–8.

[123] Bamman MM, Hill VJ, Adams GR, et al. Gender differences in resistance-training-induced myofiber hypertrophy among older adults. J Gerontol 2003;58A:108–16.

[124] Greendale GA, DeAmicis TA, Bucur A, et al. A prospective study of the effect of fracture on measured physical performance: results from the MacArthur Study (MAC). J Am Geriatr Soc 2000;48:546–9.

[125] Dai YT, Huang GS, Yang RS, et al. Functional recovery after hip fracture: six months' follow-up of patients in a multidisciplinary rehabilitation program. J Formos Med Assoc 2002;101(12):846–53.

[126] Resnick B. Efficacy beliefs in geriatric rehabilitation. J Gerontol Nurs 1998;24(7):34–44.

[127] Resnick B, Daly MP. The effect of cognitive status on outcomes following rehabilitation. Fam Med 1997;29(6):400–5.

[128] Sherrington C, Lord SR. Home exercise to improve strength and walking velocity after hip fracture: a randomized controlled trial. Arch Phys Med Rehabil 1997;78:208–12.

[129] Henderson SA, Finlay OE, Murphy N, et al. Benefits of an exercise class for elderly women following hip surgery. Ulster Med J 1992;61(2): 144–50.

[130] Baker PA, Evans OM, Lee C. Treadmill gait retraining following fractured neck-of-femur. Arch Phys Med Rehabil 1991;72(9):649–52.

[131] Lieberman D. Rehabilitation following hip fracture surgery: a comparative study of females and males. Disabil Rehabil 2004;26(2):85–90.

[132] National Blueprint. Increasing physical activity among adults age 50 and older; 2000. Available at: http://www.rwjf.org/publications/otherlist.jsp.

[133] National Osteoporosis Foundation. Physician's guide to prevention and treatment of osteoporosis. Washington, DC: National Osteoporosis Foundation; 2003.

[134] National Osteoporosis Foundation. Health professional's guide to rehabilitation of the patient with osteoporosis. Washington, DC: National Osteoporosis Foundation; 2003.

[135] Honeywell M, Phillips S, Branch E III, et al. Teriparatide for osteoporosis: a clinical review. Drug Forecast 2003;28(11):713–6.

[136] Orwoll ES, Scheele WH, Paul S, et al. The effect of teriparatide (human parathyroid hormone (1–34)) therapy on bone density in men with osteoporosis. J Bone Miner Res 2003;18:9–17.

[137] Orwig DL, Wehren LE, Yu-Yahiro JA, et al. Treatment of osteoporosis following a hip fracture: sending results of bone densitometry to primary care physicians does not increase use of pharmacologic therapy [abstract F382]. J Bone Miner Res 2001; 16(Suppl 1):S220.

[138] Juby AG, De Geus-Wenceslau CM. Evaluation of osteoporosis treatment in seniors after hip fracture. Osteoporos Int 2002;13:205–10.

[139] Solomon DH, Finkelstein JS, Katz JN, et al. Underuse of osteoporosis medications in elderly patients with fractures. Am J Med 2003;115:398–400.

[140] Simonelli C, Chen Y-T, Morancey J, et al. Evaluation and management of osteoporosis following hospitalization for low-impact fracture. J Gen Intern Med 2003;18:17–22.

[141] Kamel HK, Duthie EH. The underuse of therapy in the secondary prevention of hip fractures. Drugs Aging 2002;19:1–10.

[142] Harrington JT, Broy SB, Derosa AM, et al. Hip fracture patients are not treated for osteoporosis: a call to action. Arthritis Rheum 2002;47:651–4.

[143] Gardner MJ, Flik K, Mooar PA, et al. Improvement in the undertreatment of osteoporosis following hip fracture. J Bone Joint Surg Am 2002;84A:1342–8.

[144] Follin S, Black J, McDermott M. Lack of diagnosis and treatment of osteoporosis in men and women after hip fracture. Pharmacotherapy 2003;23:190–8.

[145] Bahl S, Coates PS, Greenspan SL. The management of osteoporosis following hip fracture: have we improved our care? Osteoporos Int 2003; 14(11):884–8.

[146] Jachna CM, Whittle J, Lukert B, et al. Effect of hospitalist consultation on treatment of osteoporosis in hip fracture patients. Osteoporos Int 2003; 14(8):665–71.

[147] Black DM, Steinbuch M, Palermo L, et al. An assessment tool for predicting fracture risk in postmenopausal women. Osteoporos Int 2001;12: 519–28.

[148] Cree MW, Juby AG, Carriere KC. Mortality and morbidity associated with osteoporosis drug treatment following hip fracture. Osteoporos Int 2003; 14(9):722–7.

[149] Bellantonio S, Fortinsky RH, Prestwood K. How well are community-living women treated for osteoporosis after hip fractures? J Am Geriatr Soc 2001;49:1197–204.

[150] Kiebzak GM, Beinart GA, Perser K, et al. Undertreatment of osteoporosis in men with hip fracture. Arch Intern Med 2002;162(19):2217–22.

[151] Folstein MF, Folstein SE, McHugh PR. "Minimental State": a practical method for grading the cognitive state of patients for the clinician. J Psychiatr Res 1975;12:189–98.

[152] Charlson M, Szatrowski TP, Peterson J, et al. Validation of a combined comorbidity index. J Clin Epidemiol 1994;47(11):1245–51.

Sexual Dimorphism in Stroke
Mary Ann Keenan, MD

The Neuro-Orthopaedics Program, Department of Orthopaedic Surgery, The University of Pennsylvania, 3400 Spruce Street, 2 Silverstein, Philadelphia, PA 19104, USA

Each year, 700,000 people in the United States suffer a stroke. Approximately 500,000 are first attacks and 200,000 are recurrent strokes [1]. In 2002, stroke killed 162,672 people; 61% of these stroke fatalities were women. Approximately 50% of stroke deaths occur before the person reaches the hospital.

More than half of stroke victims survive and have an average life expectancy of greater than 8 years [2,3]. Stroke is a leading cause of serious, long-term disability. Many of the long-term problems that are faced by stroke survivors are musculoskeletal in nature [4,5]. Among ischemic stroke survivors who were at least 65 years old, the following disabilities were observed at 6 months after stroke [6]:

Fifty percent had hemiparesis.
Thirty percent were unable to walk without assistance.
Twenty-six percent were dependent in activities of daily living (ADLs).
Nineteen percent had aphasia.
Thirty-five percent had depressive symptoms.
Twenty-six percent were institutionalized in a nursing home.

Although much is known about the long-term outcome of stroke survivors in terms of mortality and disability, there has been little research on the patient-centered outcome of health-related quality of life (HRQoL) [2]. There are limited natural history data on HRQoL beyond 2 years after stroke and no data on those factors that are present at stroke onset that predict HRQoL beyond 2 years after stroke. Most stroke survivors have the potential for significant function and useful lives if they receive the benefits of rehabilitation and reconstructive surgery for residual limb deformities.

Susceptibility to stroke

The E65K polymorphism in the Ca^{2+}-dependent K^+ (BK) channel, a key element in the control of arterial tone, has been associated with low prevalence of diastolic hypertension. Senti and colleagues [7] reported on the modulatory effect of sex and age on the association of the E65K polymorphism with low prevalence of diastolic hypertension and the protective role of E65K polymorphism against cardiovascular disease. They analyzed the genotype frequency of the E65K polymorphism in 3924 participants who were selected randomly in two cross-sectional studies. They performed a 5-year follow-up of the study group to determine whether cardiovascular events had occurred. Multivariate regression analyses showed that increasing age increased the protective effect of the K allele against moderate-to-severe diastolic hypertension in the overall group of participants. This study provides the first genetic evidence for the different impact of the BK channel in the control of human blood pressure in men and women, with particular relevance in older women. It also highlights the E65K polymorphism as one of the strongest genetic factors associated, thus far, with protection against myocardial infarction and stroke.

Little information is available on sex differences among younger adult stroke patients. Hochner-Celnikier and colleagues [8] performed a study to scrutinize differences in mortality, principal risk factors, and outcome measures among patients aged 45 to 65 years who had acute stroke. This was a retrospective study of 114 women and 190 men aged 45 to 65 years, who were hospitalized

E-mail address: maryann.keenan@uphs.upenn.edu

from 1990 to 1998 with confirmed cerebrovascular accident. No sex differences were observed in clinical presentation or imaging studies. The mortality among women was higher than among men (13.2% versus 5.8%). There was a significant sex gap in comorbidity with diabetes, hypertension, and hypercholesterolemia (29.1% of women versus 14.3% of men). More men had a history of ischemic heart disease (35.8% versus 21.9%), smoking (43.9% versus 16.4%), and alcohol consumption (6.9% versus 0.9%). The concomitance of multiple risk factors may have contributed to the observed higher mortality in the women. The use of rehabilitative services was similar between the sexes.

Fang and colleagues examined data from a large contemporary cohort of patients to address whether women are at higher risk for thromboembolism in the setting of atrial fibrillation (AF). They prospectively evaluated 13,559 adults who had AF, and recorded data on patients' clinical characteristics and the occurrence of incident hospitalizations for ischemic stroke, peripheral embolism, and major hemorrhagic events through searching validated computerized databases and medical record review. After multivariable analysis, women had higher annual rates of thromboembolism while off warfarin than did men. There was no significant difference by sex in the 30-day mortality after thromboembolism (23% for both). Warfarin use was associated with significantly lower adjusted thromboembolism rates for men and women. Men and women also had similar annual rates of major hemorrhage with warfarin use.

Diagnostic evaluation

Sixty-two percent of all stroke deaths in the United States occur in women. Smith and colleagues [9] compared diagnostic evaluations by sex in patients who had ischemic stroke in a biethnic, population-based study. The investigators selected a random sample of patients who had ischemic stroke identified between 2000 and 2002 by BASIC (Brain Attack Surveillance in Corpus Christi Project). Sex differences in the use of stroke diagnostic tests were evaluated. The study population consisted of 161 men and 220 women. The median age was 74.3 years; 71% of men and 62% of women received any carotid artery evaluation. The difference in rates of brain MRI was less, with 43% of the men and 41% of the women studied. Fifty-seven percent of the men and 48% of the women were assessed with echocardiography, and 90% of the men and 86% of the women were evaluated with EKG. Multivariable logistic models found that women were less likely to undergo echocardiography (odds ratio [OR] 0.64; 95% CI, 0.42–0.98) and carotid evaluation (OR 0.57; 95% CI, 0.36–0.91). There was no association of ischemic stroke subtype and sex to explain these results ($P = .76$). Because women were less likely to have been evaluated for heart valve disease or carotid artery occlusion, diagnosis of their disease is likely to be delayed, which leads to increased morbidity and mortality.

Preventative treatment

Dick and colleagues [10] from the Medical University Vienna investigated sex-related differences in vascular outcome and mortality of asymptomatic patients who had high-grade internal carotid artery (ICA) stenosis. They enrolled 525 consecutive patients (325 men with a median age of 72 years and 200 women with a median age of 75 years) from a single-center registry who initially were treated conservatively with respect to a neurologically asymptomatic ICA stenosis of at least 70%. Their patients were followed for a median of 38 months for major adverse cardiovascular, cerebral, and peripheral vascular events, vascular mortality, and all-cause mortality. Adjusted hazard ratios for MACE (combined end point, including myocardial infarction, stroke, [partial] limb amputation, and death), vascular mortality, and all-cause mortality for men were 1.96 ($P = .016$), 2.48 ($P < .001$), and 1.70 ($P = .007$), respectively, as compared with women, irrespective of age, vascular risk factors, comorbidities, and the individual risk status estimated by the American Society of Anesthesiologists score. Their conclusions were that men who have high-grade carotid artery stenosis are at a considerably higher risk for poor outcome than are their female counterparts. In particular, the risk for fatal vascular events is increased substantially in men. This study considered all cardiovascular events and did not look specifically at stroke as an outcome. Therefore, this study did not evaluate or explain the higher mortality of stroke in women.

Treatment of acute stroke

Coagulation and fibrinolysis differ between sexes. Previous reports suggest that women achieve better outcome than do men after intravenous thrombolysis for ischemic stroke. These findings prompted Savitz and colleagues [11] to investigate possible sex differences in arterial recanalization after intravenous tissue plasminogen activator (IV tPA). They identified 100 consecutive patients who presented with acute ischemic stroke and received IV tPA within 6 hours of onset. Only patients who had large artery anterior circulation strokes, as determined by MRI/MR angiography or CT/CT angiography before treatment, and who had follow-up vascular study within 72 hours after treatment, were included. Thirty-nine patients met the study criteria (22 men and 17 women). The recanalization rate was significantly higher in women (94% versus 59%; $P = .02$). This difference remained statistically significant after excluding patients whose strokes were attributed to ICA occlusive lesions, and when the analysis was limited to those patients who were treated within 3 hours of stroke onset. All other confounding variables did not differ significantly between the sexes.

Women are less likely than are men to receive some stroke care interventions. It is not known whether sex differences in patient preferences explain any of the variations in the delivery of stroke care. Researchers from the University of Toronto provided self-administered surveys to patients who did and did not have a history of cerebrovascular disease who were recruited from stroke, vascular, and general internal medicine ambulatory clinics [12]. The surveys described hypothetic scenarios. The participants were asked if they would accept therapy with thrombolysis for acute ischemic stroke or carotid endarterectomy for secondary stroke prevention. The survey also included questions on sociodemographic factors and decision-making preferences. A total of 586 patients (45% women) completed the survey. Women were less likely to accept thrombolysis (79% versus 86%, $P = .014$), even after adjustment for other factors. Women and men were equally likely to accept carotid endarterectomy (82% versus 84%, $P = .502$). Women were less confident in their decisions, were more averse to risk, and would have preferred more information to assist them in their decision making. This study indicates that health care providers need to be aware that women may be more concerned about risks and may require more information before they make a decision.

Anand and colleagues [13] performed a similar study to determine if sex and sex differences in the management of acute coronary syndromes (ACSs) were associated with differences in prognosis after ACSs. They analyzed data from the Clopidogrel in Unstable Angina to Prevent Recurrent Events (CURE) trial. The trial enrolled 4836 women and 7726 men who had ACSs. Patients were classified into risk strata using the Thrombolysis in Myocardial Infarction (TIMI) score. Their conclusions were that high-risk women who have ACSs undergo less coronary angiography, angioplasty, and coronary artery bypass graft surgery compared with men. Women who have ACSs do not have a higher incidence of cardiovascular death, recurrent myocardial infarction, or stroke. Nevertheless, women do exhibit an increased rate of refractory ischemia and rehospitalization.

Prevention of recurrent stroke

Results of aspirin teamed with the antiplatelet drug Persantine (dipyridamole) versus aspirin alone for the prevention of vascular events after ischemic stroke have been conflicting and inconsistent. A 79-hospital, 14-country study, known as the European/Australasian Stroke Prevention in Reversible Ischemia Trial (ESPIRIT), was coordinated by researchers at the University Medical Center Utrecht in an effort to examine this issue [14,15]. In a randomized controlled trial, 1363 patients were assigned to 30 mg to 325 mg of aspirin daily and Persantine, 200 mg twice daily, whereas 1376 patients were given aspirin alone within 6 months of a transient ischemic attack (TIA) or minor stroke. The mean length of follow-up was 3.5 years. Median aspirin dosage was 75 mg in both treatment groups (30–325 mg); extended-release Persantine was used by 83% (1131 patients) of the patients who received the combination regimen. The primary outcome events included the composite of death from all vascular causes, nonfatal stroke, nonfatal myocardial infarction, or major bleeding complication, whichever happened first. The patients who received dual therapy had fewer circulatory events than did those who took aspirin alone. A primary event occurred in 173 patients (13%) who were on combined treatment, versus 216 patients (16%) who were on aspirin alone. The absolute risk reduction was 1.0% per year for combined treatment. A drawback of the dual therapy was that patients discontinued the medications more often than did those who were taking aspirin alone (470 versus 184 patients). Thirty-four percent discontinued treatment because of adverse effects, including 26% (123 patients) who reported headache as one of the reasons. Of patients who were allocated to aspirin therapy, 184 patients (13%) discontinued their medication primarily because of medical reasons, including new TIA, stroke, or an indication for oral anticoagulant therapy. Subgroup analysis indicated that the results were valid for men and women. There was a greater effect seen in men, but no further investigation of this difference was performed.

Burden of care after stroke

An increasing number of people is forced to enter the role of informal caretaker as the population ages. Despite the increase in the need for caregivers and the importance of providing care, there is little empiric research examining how men and women approach and cope with providing care. Researchers at the University of Florida in Gainesville examined stroke survivors and their care providers to assess possible sex differences in the impact of caretaking on caretakers and care recipients [16]. Their results indicate no significant difference in patient well-being based on the sex of the caregiver. Some measures indicated that men have advantages as caregivers.

Appelros and colleagues [17] from Orebro University Hospital in Sweden performed a similar study in which they examined the living setting and need for ADL assistance before and 1 year after a first-ever stroke, with special focus on sex differences in 377 people. Before the stroke, 48 patients (13%) lived in special housing (assisted living or nursing homes), and 1 year after the stroke, 50 of the survivors (20%) required special accommodations. Before the onset of the stroke, only 80 (21%) of the patients needed help with their personal ADLs. One year after the stroke, 90 (36%) needed ADL assistance. The increased need was fulfilled by relatives. Female spouses more often helped their male counterparts, and they tended to accept a heavier burden.

Researchers from The Sahlgrenska Academy at Goteborg University in Sweden performed a study to evaluate formal care and the situation of informal caregivers from a gender perspective [18]. The study targeted 147 elderly people (94 women, 53 men) who were living in their own homes 12 months following an acute stroke. The median age of the women was 81 years and the median age of then men was 80 years. Statistically significant sex differences in living conditions were seen. Eighty percent of the women were living alone, whereas only 28% of the men lived alone. The informal care given far exceeded that provided by formal community services. Some type of informal care was provided to 65% of these elderly people, whereas only 44% received formal care from the community. A sex difference in daily informal personal care was noted, with 24% of men and 16% of women receiving assistance. Formal care was provided by the community significantly more frequently to women (56%), most of whom were living alone, than to men (23%), who were less likely to be living alone. The women more frequently had community-based help with housecleaning and they also more frequently received help with personal care. Most of the caregivers were elderly women, and preventive intervention measures should be developed to enable them to manage their everyday lives.

In the United States there are many adults living with disabilities. Little is known about how sex differences affect stroke risk factors in this population. A descriptive study was performed by researchers from Villanova University College of Nursing to determine whether men and women living with disabilities differed in self-reported rates of stroke risk factors [19]. There were 146 participants, 54% were women, and the mean age was 58 years. The primary instrument was the Stroke Risk Screening tool. Stroke risk factors that differed significantly by sex included the incidence of hypertension (48% of men versus 32% of women), current smoking (30% of men versus 4% of women), history of heart disease (13% of men versus 1% of women), daily consumption of alcohol (10% of men versus 1% of women), and use of illicit drugs (10% of men versus 0% of women).

Neurologic impairment

Infarction of the cerebral cortex in the region of the brain that is supplied by the middle cerebral artery or one of its branches is responsible for stroke most commonly. Although the middle cerebral artery supplies the area of the cerebral cortex that is responsible for hand function, the anterior cerebral artery supplies the area that is responsible for lower extremity motion. The typical clinical picture following middle cerebral artery stroke is contralateral hemianesthesia, homonymous hemianopia, and spastic hemiplegia with more paralysis in the upper extremity than in the lower extremity [20–22]. Because hand function requires precise motor control, even for activities with assistive equipment, the prognosis for the functional use of the hand and arm is considerably worse than for the leg. Return of even gross motor control in the lower extremity may be sufficient for walking.

Infarction in the region of the anterior cerebral artery causes paralysis and sensory loss of the opposite lower limb, and, to a lesser degree, the arm. Patients who have cerebral arteriosclerosis and suffer repeated bilateral infarctions are likely to have severe cognitive impairment that limits their general ability to function, even when motor function is good.

Neurologic recovery

After stroke, motor recovery follows a fairly typical pattern for men and women. The size of the lesion and the amount of collateral circulation determine the amount of permanent damage. Most recovery occurs within 6 months, although functional improvement may continue as the patient receives further sensorimotor re-education and learns to cope with disability [23].

Initially after a stroke, the limbs are completely flaccid. Over the next few weeks, muscle tone and spasticity gradually increase in the adductor muscles of the shoulder and in the flexor muscles of the elbow, wrist, and fingers. Spasticity also develops in the lower extremity muscles. Most commonly, there is an extensor pattern of spasticity in the leg, which is characterized by hip adduction, knee extension, and equinovarus deformities of the foot and ankle. In some cases, however, a flexion pattern of spasticity occurs, which is characterized by hip and knee flexion.

Whether the patient recovers the ability to move one joint independently of the others (selective movement) depends on the extent of the cerebral cortical damage [20,24–28]. Dependence on the more neurologically primitive patterned movement (synergy) decreases as selective control improves. The extent to which motor impairment restricts function varies in the upper and lower extremities. Patterned or synergistic movement is not of functional value in the upper extremity. Synergistic movement may be useful in the lower extremity, where the patient uses the flexion synergy to advance the limb forward and the mass extension synergy for limb stability during standing.

The final processes in sensory perception occur in the cerebral cortex, where basic sensory information is integrated to complex sensory phenomena, such as vision, proprioception, and perception of spatial relationships, shape, and texture. Patients who have severe parietal dysfunction and sensory loss may lack sufficient perception of space and awareness of the involved segment of their body to ambulate. Patients who have severe perceptual loss may lack balance to sit, stand, or walk. A visual field deficit further interferes with limb use, and it may cause patients to be unaware of their own limbs.

Treatment of chronic mobility and functional impairment following stroke

Despite sex differences in the incidence, etiology, and response to treatment of acute stroke, differences in the incidence and treatment of mobility problems that result from the musculoskeletal impairments following stroke have not been evaluated between men and women. This is an area that needs evaluation.

Summary

Stroke is a leading cause of death and serious, long-term disability. Studies evaluating differences between men and women are seriously lacking. Significant differences exist between men and women in terms of risk factors and susceptibility to stroke. Women are less likely to have diagnostic studies performed to evaluate their risk for stroke, and have a higher mortality following acute stroke. Women, however, have a higher rate of arterial recanalization after IV tPA used for the treatment of acute stroke. The data that compare the effectiveness of treatments for prevention of recurrent stroke between men and women is sparse. Women are more likely to be impacted negatively by the burden of care for a loved one who had a stroke. There have not been any studies that compare the results of treatment of musculoskeletal impairments in men and women after stroke.

References

[1] AHA/ASA. Heart disease and stroke statistics—2006 update. Dallas (TX): American Heart Association/American Stroke Association; 2006.
[2] Haacke C, Spottke A, Siebert U, et al. Long-term outcome after stroke: evaluating health-related

quality of life using utility measurements. Stroke 2006;37(1):193–8.
[3] Hankey G. Long-term outcome after ischaemic stroke/transient ischaemic attack. Cerebrovasc Dis 2003;16(Suppl 1):14–9.
[4] Keenan MAE. Lower limb surgery for stroke patients. In: Condie E, Campbell J, Martina J, editors. Report of a Consensus Conference on the Orthotic Management of Stroke Patients. Copenhagen (Denmark): International Society for Prosthetics and Orthotics, Borgervaenget; 2004. p. 152–61.
[5] Keenan MAE. Orthopaedic management of upper extremity dysfunction following stroke. In: Condie E, Campbell J, Martina J, editors. Report of a Consensus Conference on the Orthotic Management of Stroke Patients. Copenhagen (Denmark): International Society for Prosthetics and Orthotics, Borgervaenget; 2004. p. 238–51.
[6] Hurst W. The heart, arteries and veins. New York: McGraw-Hill; 2002.
[7] Senti M, Fernandez-Fernandez JM, Tomas M, et al. Protective effect of the KCNMB1 E65K genetic polymorphism against diastolic hypertension in aging women and its relevance to cardiovascular risk. Circ Res 2005;97(12):1360–5.
[8] Hochner-Celnikier D, Manor O, Garbi B, et al. Gender gap in cerebrovascular accidents: comparison of the extent, severity, and risk factors in men and women aged 45–65. Int J Fertil Womens Med 2005;50(3):122–8.
[9] Smith MA, Lisabeth LD, Brown DL, et al. Gender comparisons of diagnostic evaluation for ischemic stroke patients. Neurology 2005;65(6):855–8.
[10] Dick P, Sherif C, Sabeti S, et al. Gender differences in outcome of conservatively treated patients with asymptomatic high-grade carotid stenosis. Stroke 2005;36(6):1178–83.
[11] Savitz SI, Schlaug G, Caplan L, et al. Arterial occlusive lesions recanalize more frequently in women than in men after intravenous tissue plasminogen activator administration for acute stroke. Stroke 2005; 36(7):1447–51.
[12] Kapral MK, Devon J, Winter AL, et al. Gender differences in stroke care decision-making. Med Care 2006;44(1):70–80.
[13] Anand SS, Xie CC, Mehta S, et al. Differences in the management and prognosis of women and men who suffer from acute coronary syndromes. J Am Coll Cardiol 2005;46:1845–51.
[14] Halkes PH, van Gijn J, Kappelle LJ, et al. Aspirin plus dipyridamole versus aspirin alone after cerebral ischaemia of arterial origin (ESPRIT): randomised controlled trial. Lancet 2006; 367(9523):1665–73.
[15] Norrving B. Dipyridamole with aspirin for secondary stroke prevention. Lancet 2006;367(9523):1638–9.
[16] Tiegs TJ, Heesacker M, Ketterson TU, et al. Coping by stroke caregivers: sex similarities and differences. Top Stroke Rehabil 2006;13(1):52–62.
[17] Appelros P, Nydevik I, Terent A. Living setting and utilisation of ADL assistance one year after a stroke with special reference to gender differences. Disabil Rehabil 2006;28(1):43–9.
[18] Gosman-Hedstrom G, Claesson L. Gender perspective on informal care for elderly people one year after acute stroke. Aging Clin Exp Res 2005;17(6):479–85.
[19] Hinkle JL, Smith R, Revere K. A comparison of stroke risk factors between men and women with disabilities. Rehabil Nurs 2006;31(2):70–7.
[20] Botte MJ, Abrams RA, Keenan MAE, et al. Limb rehabilitation in stroke patients. The Journal of Musculoskeletal Medicine 1992;9:66–78.
[21] Botte MJ, Keenan MAE. Brain injury and stroke. In: Gelberman RH, editor. Operative nerve repair and reconstruction. Philadelphia: Lippincott Publishers; 1991. p. 1413–51.
[22] Botte MJ, Keenan MAE, Jordan C. Stroke. In: Nickel VL, Botte MJ, editors. Orthopaedic rehabilitation. New York: Churchill Livingstone; 1992. p. 337–60.
[23] Weimar C, Konig IR, Diener HC. Predicting functional outcome and survival after acute ischemic stroke. J Neurol 2002;249(7):888–95.
[24] Keenan MA, Haider TT, Stone LR. Dynamic electromyography to assess elbow spasticity. J Hand Surg [Am] 1990;15(4):607–14.
[25] Keenan MA, Romanelli RR, Lunsford BR. The use of dynamic electromyography to evaluate motor control in the hands of adults who have spasticity caused by brain injury. J Bone Joint Surg Am 1989;71(1):120–6.
[26] Keenan MAE, Perry J. Evaluation of upper extremity motor control in spastic brain-injured patients using dynamic electromyography. J Head Trauma Rehabil 1990;5(4):13–22.
[27] Kozin SH, Keenan MA. Using dynamic electromyography to guide surgical treatment of the spastic upper extremity in the brain-injured patient. Clin Orthop 1993;288:109–17.
[28] Pinzur M. Dynamic electromyography in functional surgery for upper limb spasticity. Clin Orthop 1993; 288:118–21.

Sex Differences in Autoimmune Disease
Michael D. Lockshin, MD

Barbara Volcker Center for Women and Rheumatic Disease, Mary Kirkland Center for Lupus Research,
Joan and Sanford Weill Medical College of Cornell University, Hospital for Special Surgery,
535 East 70th Street, New York, NY 10021, USA

Some human autoimmune rheumatic diseases affect more women than men, but disease severity does not differ between the sexes. Quantities of sex discrepancies differ among the autoimmune diseases. Female predominance occurs in thyroid diseases (Hashimoto's, Graves'), some rheumatic diseases (lupus, rheumatoid arthritis, scleroderma, Sjögren's), and some hepatic diseases (autoimmune hepatitis, primary biliary cirrhosis). Male predominance characterizes other rheumatic diseases (ankylosing spondylitis, Reiter syndrome, and vasculitis) and an immunologically driven nephritis (Goodpasture's syndrome). Other autoimmune diseases—juvenile-onset diabetes, inflammatory bowel disease, and idiopathic thrombocytopenic purpura—are sex neutral [1–3]. Some of the basic facts are disputed, however. Published female/male (F/M) ratios vary tenfold (from 10–50) for Hashimoto's disease, sevenfold (from 1.5–10) for multiple sclerosis, fivefold (from 0.2–1) for Goodpasture's disease, fourfold (from 3–12) for scleroderma, and threefold (from 7–20) for lupus) (Table 1) [4–6].

Pathogenesis

Most sex-discrepant autoimmune human illnesses are equally severe in both sexes [7]. In animal autoimmune disease, when F/M ratios are high, disease severity is as well. Many authors attribute sex discrepancy to sex hormones. This

attribution focuses on estrogen's effects on in vitro immunity, in vivo amelioration of experimental disease by female castration or worsening by male castration/estrogen supplement, or human case reports in which castration or pharmacologic intervention has altered the severity of a clinical course. If immune response is inherently different between men and women, however, sex-discrepant responses to vaccination, infection, and immunomodulation should occur.

A test of the hypothesis that hormones modulate immune response in vivo is to examine sexual dimorphism in the response to vaccination or infection. The few studies that do look at sex differences after vaccination show modestly higher antibody titers in women but no other sex differences in clinical protection by, or adverse reactions to, vaccination. Most studies find male and female responses identical, an exception being that arthritic reactions to rubella are more common in women. Viral and bacterial infections affect men and women equally. Therapeutically administered cytokines induce autoimmune rheumatic symptoms equally in men and women [8], but more women who receive the anti–tumor necrosis factor-α agent infliximab for Crohn's disease develop antinuclear antibodies [9]. Although antibody titers tend to be higher in women, differences in clinically important immune response do not account for the high F/M ratios of rheumatic disease.

In many "experiments of nature" environmental exposures induce autoimmune disease. More men take drugs that induce lupus; unlike idiopathic lupus, drug-induced lupus is male predominant [10]; more men are exposed to silica inducers of scleroderma-like disease, also male

---

This article supported by the Barbara Volcker Center for Women and Rheumatic Disease and the Mary Kirkland Center for Lupus Research.

E-mail address: LockshinM@hss.edu

Table 1
Sex ratios of autoimmune diseases

F/M ratio	Diseases
9:1	Sjögren
	Hashimoto
	Graves'
	Systemic lupus erythematosus
2–3:1	Myasthenia gravis
	Multiple sclerosis
	Rheumatoid arthritis
~1:1	Autoimmune hemolytic anemia
	Idiopathic thrombocytopenic purpura
	Type I diabetes
	Vitiligo
	Pemphigus
<1:1	Goodpasture
	Ankylosing spondylitis

predominant; more women were exposed to the contaminated cooking oil that caused a scleroderma-like illness in Spain, which was female predominant [11]. More women took contaminated L-tryptophan; the resulting epidemic of eosinophilia-myalgia syndrome was female predominant [12]. In acute Lyme disease, boys are more often affected than girls because of their greater exposure in outdoor play to infected ticks; when incidence of chronic Lyme disease is adjusted for acute disease exposure, there is no sex discrepancy [13]. In many experiments of nature, the main determinant of F/M ratios is rate of exposure.

Serologic lupus begins decades before the first symptoms appear [14,15]. To identify a sex-discrepant environmental cause requires inquiry about exposure decades before first symptoms. Potential differences include exposures caused by different recreational experiences, processing infecting organisms caused by different routes of exposure (eg, menstruation and intercourse render women susceptible to infection in ways that men are not), vulnerable periods (eg, the high attack rate of malaria in the postpartum period is an example) [16], and threshold differences in immune responses.

Judging from available clinical experience, sex hormones are at best weak explanations for high F/M ratios. Population studies on effects of hormone therapy show either no or small increases of incidence of rheumatoid arthritis and lupus in patients taking these drugs. Estrogen replacement therapy, oral contraceptives, and ovulation induction probably do not worsen lupus [17,18]. Although synoviocyte estrogen receptors may be target organs in rheumatoid arthritis [19], these receptors are present in synovium of patients with chronic Lyme disease and ankylosing spondylitis, which are not female predominant. Androgens have no apparent role in the male predominance of ankylosing spondylitis [20].

The effect of pregnancy on autoimmune disease is variable. Rheumatoid arthritis and multiple sclerosis remit during pregnancy [21]. Lupus does not or only slightly worsens during pregnancy [22]. Ankylosing spondylitis worsens. Pregnancy changes in disease severity can be caused by placental or maternal hormone, increased circulation, increased fluid volume, metabolic rate, hemodilution, circulating fetal cells, or other factors. A threshold mechanism could explain an increase in incidence without a corresponding increase in severity. An animal model suggests this possibility: estrogen may permissively allow survival of forbidden autoimmune clones [23].

Hormones might influence F/M ratios by affecting nonimmune cells; for instance, hormone effects on endothelium might be critical for disease initiation. An unknown sex difference related to ovulation or menstruation cytokines, vascular rheology, or a biologic clock might be responsible for different disease experiences of the two sexes. In some mouse strains, healing is sexually dimorphic, the dimorphism being under estrogen control [24]. At the single-cell level, male and female cells in culture are strikingly different (Z. Zakeri, PhD, personal communication, 2004; J. Huard, Phd, personal communication, 2004).

Evidence for genetic control of autoimmunity is strong for spondyloarthropathy, rheumatoid arthritis, and lupus. Sex-discrepant human leukocyte antigen–associated effects [25] and genes on X and Y chromosomes are possible causes of sex discrepancy. Although ankylosing spondylitis has no X-chromosome susceptibility locus [26], CD40 ligand, some interferon-related genes, and other immunologically relevant genes are on the X or Y chromosome. Evidence that supports or refutes a chromosomal explanation for sex discrepancy is as follows: Susceptibility to lupus resides on the Y chromosome in the male-predominant BXSB mouse model [27]. Differences in imprinting or differential X-inactivation have been sought but not found [28,29]. Skewed X-inactivation in the thymus may lead to inadequate thymic deletion and loss of T-cell tolerance [30]. Mutation of a tissue/developmental stage-specific proteasome product is sex discrepant in a mouse model of diabetes. Sex dimorphism of T-cell trafficking may be caused by sex-determined cell surface markers [9]. Sex chromosomal differences are possible reasons for the sex discrepancy of autoimmune disease.

Immunization, in-breeding, and transgenic and gene knockout animal models of autoimmune disease give mixed messages about causes of sex discrepancy. In strains of mice and rats that are susceptible to experimental thyroiditis, estrogen increases antithyroid antibody titer but not histologic thyroiditis, whereas the severity of induced mouse thyroiditis varies with iodide content of diet and types of chow. In this model, genetic and extrinsic factors influence experimental thyroiditis incidence more than hormones.

Although the (NZB x NZW)F1 mouse model of lupus shows high female incidence and severity, the MRL lpr/lpr model is sex neutral and the BXSB model is male predominant [31]. Castration/replacement experiments in (NZB x NZW)F1 mice demonstrate estrogen enhancement and testosterone suppression of spontaneous disease severity and incidence. Genetic susceptibility is linked to major histocompatibility locus (MHC) and genes that control complement and apoptosis. Like its human counterpart, lupus in mice develops in young adulthood, which implies that incubation, maturation, or cumulative damage is required for disease expression.

Male and female mice in germ-free environments are equally affected by lupus, but germ-free female mice develop higher autoantibody levels. Germ-free, antigen-free animals have less frequent disease than do germ-free or conventionally raised animals, which indicates environmental contribution to illness and leaves open the possibility that differential exposure causes sex discrepancy in humans [32]. The p21 knockout and the DNAse 1 knockout mouse lupus models show slightly higher autoantibody levels in female mice. Inexplicably, glomerulonephritis is much worse in female p21 knockout mice but equals that of male mice in DNAse 1 knockouts [33,34]. The human leukocyte antigen B27 gene transgenically expressed in rats induces a phenotype with features of psoriasis and ankylosing spondylitis. In a germ-free environment, the spondylitis does not occur. Introduction of specific gastrointestinal pathogens to the germ-free animal induces spondylitis [35]. Male predominance is true of this model, as it is of the human disease, but the reasons are unknown. In these animal models of autoimmune disease, genetic, hormone, life stage, and environmental factors are all relevant to disease causation. No consistent cause for sex discrepancy appears.

Men and women differ in ways that are not easily explained by hormones, chromosomes, or specific genes, such as body size and the monthly (chronobiologic) cycle in women. Most female predominant diseases cluster in the young-adult years, whereas autoimmune diseases that affect younger or older patients are more evenly divided between the sexes. Characteristics of young adulthood that may explain female predominance include mode of sexual intercourse, pregnancy, chronobiology, nonhormonal effects of menstrual cycles, vascular responses, and as-yet unknown other variables. The large quantity and long duration of circulating fetal cells in scleroderma and other autoimmune disease patients [36] and the finding of pregnancy-created chimerism in sites of autoimmune disease [37] suggest a profound new biologic difference between men and women, the implications of which are unknown.

Explanations for sex discrepancy

In non-autoimmune human illnesses, the most striking differences of incidence occur when exposures to infectious agents or toxins differ between the sexes. If infections or toxins induce autoimmune disease, differences in exposure (perhaps decades before onset of clinical illness) remain as plausible explanations for the sex differences. If gonadal hormones play a role, they likely do so through a threshold or permissive mechanism rather than through immunomodulation. Differences related to X-inactivation, imprinting, X or Y chromosome genetic modulators, and intrauterine influences remain as alternate, theoretical explanations for sex differences of incidence. The epidemiology of some autoimmune diseases—young, female—suggests that an explanation for female predominance lies in exposure, vulnerable periods, or thresholds rather than in the immune response itself. These topics remain to be explored.

Summary

Many, but not all, autoimmune diseases primarily affect women. Female/male ratios range from 10:1 to 1:3. In humans, severity of illness does not differ between men and women. Men and women respond similarly to infection and vaccination, which suggests that the intrinsic differences in immune response between the sexes do not account for differences in disease frequency. In autoimmune-like illnesses caused by recognized environmental agents, sex discrepancy is usually explained by differences in exposure. Because the beginnings of autoimmune diseases can be identified decades before clinical illness, a causative factor also would have to be sought decades before illness. Endogenous hormones are not a likely explanation for sex discrepancy; hormones could have an effect if the effect is a threshold rather than quantitative. X and Y chromosomal differences have not been studied in depth. Other possibilities to explain sex discrepancy include chronobiologic differences and various other biologies, such as pregnancy and menstruation, in which men differ from women.

References

[1] Shoenfeld Y, Cervera R. Innovations in autoimmunity in the last decade. In: Shoenfeld Y, editor. The decade of autoimmunity. Amsterdam: Elsevier; 1999. p. 7–18.
[2] Janeway CA, Travers P, Walport M, editors. Immunobiology. 4th edition. New York: Elsevier Science Ltd./Garland Publishing; 1999. p. 489–509.
[3] Feltkamp TEW. The mystery of autoimmune diseases. In: Shoenfeld Y, editor. The decade of autoimmunity. Amsterdam: Elsevier; 1999. p. 1–5.
[4] AARDA. Home page. Available at: http://www.aarda.org/. Accessed November 17, 2000.
[5] Rose NR, Mackay IR. The autoimmune diseases. 3rd edition. San Diego: Academic Press; 1998. p. 1–4.
[6] Beeson P. Age and sex association of 40 autoimmune diseases. Am J Med 1994;96:457–62.
[7] Weyand CM, Schmidt D, Wagner U, et al. The influence of sex on the phenotype of rheumatoid arthritis. Arthritis Rheum 1998;41:817–22.
[8] Ioannou Y, Isenberg DA. Current evidence for the induction of autoimmune rheumatic manifestations by cytokine therapy. Arthritis Rheum 2000;43:1431–42.
[9] Vermeire S, Noman M, Van Assche G, et al. Autoimmunity associated with anti-tumor necrosis factor alpha treatment in Crohn's disease: a prospective cohort study. Gastroenterology 2003;125(1):32–9.
[10] Yung R, Williams R, Johnson K, et al. Mechanisms of drug-induced lupus. III. Sex-specific differences in T cell homing may explain increased disease severity in female mice. Arthritis Rheum 1997;40:1334–43.
[11] Abaitua Borda I, Philen RM, Posada de la Paz M, et al. Toxic oil syndrome mortality: the first 13 years. Int J Epidemiol 1998;27:1057–63.
[12] Shulman LE. The eosinophilia-myalgia syndrome associated with ingestion of L-tryptophan. Arthritis Rheum 1990;33:913–7.
[13] Carlson D, Hernandez J, Bloom BJ, et al. Lack of *Borrelia burgdorferi* DNA in synovial samples from patients with antibiotic treatment-resistant Lyme arthritis. Arthritis Rheum 2000;42:2705–9.
[14] Reichlin M, Harley JB, Lockshin MD. Serologic studies of monozygotic twins with systemic lupus erythematosus. Arthritis Rheum 1992;35:457–64.
[15] Arbuckle MR, McClain MT, Rubertone MV, et al. Development of autoantibodies before the clinical onset of systemic lupus erythematosus. N Engl J Med 2003;349(16):1526–33.
[16] Diagne N, Rogier C, Sokhna CS, et al. Increased susceptibility to malaria during the early postpartum period. N Engl J Med 2000;343:598–603.
[17] Guballa N, Sammaritano L, Schwartzman S, et al. Ovulation induction and in vitro fertilization in systemic lupus erythematosus and antiphospholipid syndrome. Arthritis Rheum 2000;43:550–6.
[18] Petri M, Kim MY, Kalunian KC, et al. Combined oral contraceptives in women with systemic lupus erythematosus. N Engl J Med 2005;353:2550–8.
[19] Castagnetta L, Cutolo M, Granata OM, et al. Endocrine end-points in rheumatoid arthritis. Ann N Y Acad Sci 1999;876:180–91.
[20] Giltay EJ, van Schaardenburg D, Gooren LJ. Androgens and ankylosing spondylitis: a role in the pathogenesis? Ann N Y Acad Sci 1999;876:340–64.
[21] Nelson JL, Hughes KA, Smith AG, et al. Maternal-fetal disparity in HLA class II alloantigens and the pregnancy-induced amelioration of rheumatoid arthritis. N Engl J Med 1993;329:466–71.
[22] Lockshin MD. Does lupus flare during pregnancy? Lupus 1993;2:1–2.
[23] Bynoe MS, Grimaldi CM, Diamond B. Estrogen up-regulates Bcl-2 and blocks tolerance induction of naïve B cells. Proc Natl Acad Sci U S A 2000;97(6):2703–8.
[24] Heber-Katz E, Chen P, Clark L, et al. Regeneration in MRL mice: further genetic loci controlling the ear hole closure trait using MRL and M.m. Castaneus mice. Wound Repair Regen 2004;12(3):384–92.
[25] Lambert NC, Distler O, Muller-Ladner U, et al. HLA-DQA1 *0501 is associated with diffuse systemic sclerosis in Caucasian men. Arthritis Rheum 2000;43:2005–10.
[26] Hoyle E, Laval SH, Calin A, et al. The X-chromosome and susceptibility to ankylosing spondylitis. Arthritis Rheum 2000;43:1353–5.
[27] Schrott LM, Waters NS, Boehm GW, et al. Behavior, cortical ectopias, and autoimmunity in BXSB-Yaa and BXSB-Yaa + mice. Brain Behav Immun 1993;7(3):205–23.
[28] Stewart JJ. The female X-inactivation mosaic in systemic lupus erythematosus. Immunol Today 1998;19:352–7.
[29] Trejo V, Derom C, Vlietinck R, et al. X chromosome inactivation patterns correlate with fetal-placental

anatomy in monozygotic twin pairs: implications for immune relatedness and concordance for autoimmunity. Mol Med 1994;1:62–70.
[30] Chitnis S, Monteiro J, Glass D, et al. The role of X-chromosome inactivation in female predisposition to autoimmunity. Arthritis Res 2000;2(5):399–406.
[31] Lahita RG. Gender and age in lupus. In: Lahita RG, editor. Systemic lupus erythematosus. 3rd edition. San Diego: Academic Press; 1999. p. 131.
[32] Maldonado MA, Kakkanaiah V, MacDonald GC, et al. The role of environmental antigens in the spontaneous development of autoimmunity in MRL-lpr mice. J Immunol 1999;162:6322–30.
[33] Balomenos D, Martin-Caballero J, Garcia MI, et al. The cell cycle inhibitor p21 controls T-cell proliferation and sex-linked lupus development. Nat Med 2000;6:171–6.
[34] Napirei M, Karsunky H, Zevnik B, et al. Features of systemic lupus erythematosus in Dnase 1-deficient mice. Nat Genet 2000;25:177–81.
[35] Taurog JD, Maika SD, Satumtira N, et al. Inflammatory disease in HLA-B27 transgenic rats. Immunol Rev 1999;169:209–23.
[36] Evans PC, Lambert N, Maloney S, et al. Long-term fetal microchimerism in peripheral blood mononuclear cell subsets in healthy women and women with scleroderma. Blood 1999;93:2033–7.
[37] Khosrotehrani K, Johnson KL, Cha DH, et al. Transfer of fetal cells with multilineage potential to maternal tissue. JAMA 2004;292(1):75–80.

Sex-Based Centers of Care: A Look to the Future
Kimberly J. Templeton, MD
University of Kansas Medical Center, 3901 Rainbow Blvd, MS 3017, Kansas City, KS 66160-7387, USA

The establishment of sex-based health centers significantly antedates the clinical and basic science revolution that led to sex-based health care. Early clinics generally focused on women's reproductive health because of the morbidity and mortality associated with pregnancy and childbirth. When men's clinics first appeared, they also focused on reproductive issues.

As part of its effort to address continuing deficits in women's health care, the Office of Women's Health of the Department of Health and Human Services (DHHS) established 18 multidisciplinary, collaborative National Centers of Excellence (CoEs) in Women's Health in 1996 [1]. The CoEs were initially housed in academic institutions; later, community CoEs were introduced. The mission of the CoEs is to improve women's health from the perspectives of clinical care, research, curricular change (medical school, clinical training, research, and continuing education), community/public outreach, outcomes assessment, and leadership development of female faculty [2]. In addition, CoEs are intended to increase awareness of sex- and gender-based differences in disease.

Women's CoEs typically focus on primary care. Most physicians are in internal medicine, family practice, or obstetrics/gynecology, although some centers include other disciplines such as cardiology and psychiatry/psychology. Studies indicate that women who receive care at a CoE are more likely to receive the screening examinations (ie, routine physical examination, Pap test, breast examination, mammogram, cholesterol test, and colon cancer screening) recommended by the DHHS than women who receive their care from a standard primary care practice [3–5]. The higher rate of screening may reflect the emphasis that CoEs (or the institutions in which they are located) place on prevention; however, it may also be due to the fact that the overwhelming majority of providers in these settings are women, who tend to provide more preventive services in general than male physicians [3,6,7].

The CoEs, however, seem to pay scant attention to musculoskeletal health promotion in women. In a survey of female patients regarding which quality of care benchmarks were addressed by their physicians, the musculoskeletal topic most frequently mentioned was domestic violence [8]. In addition, women receiving care at a CoE are more likely to receive counseling regarding violence than are women who receive care through a standard primary care practice [9].

In a review of patients seen at the Yale CoE, musculoskeletal conditions were second only to depression in frequency of presenting complaints [9]. Other than domestic violence, however, few musculoskeletal topics are mentioned in the reviews of CoEs. This is a glaring omission given that musculoskeletal conditions are among the most common causes of disability among older women. For example, osteoporosis affects approximately half of all women over age 50 years, and osteoporosis and related fragility fractures are two of the most common causes of disability among older women.

To date, the most significant effort by centers to promote musculoskeletal health has been in public education. The University of Maryland Women's Health Resource Center provides public information on physical fitness. The CoEs at the University of Michigan and the University of Washington provide public education on osteoporosis. In addition, osteoporosis was the first topic selected in the University of Washington CoE medical student community outreach

E-mail address: ktemplet@kumc.edu

0030-5898/06/$ - see front matter © 2006 Elsevier Inc. All rights reserved.
doi:10.1016/j.ocl.2006.09.012

orthopedic.theclinics.com

project, with students in this program disseminating information from the National Osteoporosis Foundation. The minimedical school program at the Indiana University School of Medicine included a session on fitness in 2000. In 1999, the CoE at Wake Forest included osteoporosis among six topics presented in a public education program [10].

These outreach efforts, laudable as they are, only begin to deal with patients' chief concerns. For example, Tannenbaum and Mayo [11], in a survey of community-dwelling women between ages 55 and 93 years, found that women were more likely to be concerned about preventing disability than with developing disease. The women listed mobility problems as their second priority, second only to loss of vision. The participants believed that physicians' performances in the area of preservation of mobility and fall prevention were not meeting their priorities.

On a positive note, osteoporosis may receive increased emphasis in the future. The Office of Women's Health is surveying community CoEs to assess how they coordinate health care services for women; osteoporosis is one of the areas to be addressed. Other musculoskeletal conditions that have an increased incidence or variable presentation in women, including osteoarthritis, rheumatoid arthritis, and fibromyalgia, however, are not included even though these conditions can lead to significant deterioration in quality of life for women.

CoE directors appear ready to shift their emphasis, at least in part. A 2001 survey [12] revealed that they believe that research into conditions that have an impact on quality of life is as important as research into potentially fatal conditions, recommending a paradigm shift from length of life to quality of life for older women. The top three research areas identified in this survey were cardiovascular health, breast cancer, and mental health. Ironically, despite the impact of musculoskeletal conditions on quality of life, the only musculoskeletal topic mentioned in this survey was domestic violence, and it was listed by only one center director.

But, a paradigm shift is beginning. After review of its patient population, Henrich and colleagues [9] of the Yale CoE recommended interdisciplinary teaching on a variety of musculoskeletal topics including osteoporosis, musculoskeletal complaints, and acute pain (including back pain). The Association of Professors of Gynecology and Obstetrics now includes a section on musculoskeletal conditions including osteoporosis, arthritis, inflammatory conditions, and soft-tissue injury in its medical student curriculum guide [13]. The Association of American Medical Colleges (AAMC) published its support for this curriculum in 2005, stating "providing quality health care to people of all ages clearly requires an understanding of the differences and similarities attributed to sex and gender" [14]. In addition, the AAMC has demonstrated support for increased medical student education through its Medical School Objectives Project, Contemporary Issues in Medicine: Musculoskeletal Medicine Education, published in 2005. The report of this project stated, "the attention paid to the (musculoskeletal) conditions in the usual medical school curriculum is not commensurate with the prevalence of these conditions" and described the knowledge, skills, and attitudes relevant to this area that the panel recommended all medical students should acquire before graduation.

Traditionally, women's health care was synonymous with reproductive care. The CoEs have raised the visibility of women's health from a clinical and a research standpoint. They have focused attention on sex- and gender-based differences in a variety of areas. The CoEs, however, appear to have paid little attention to such differences in the musculoskeletal arena. The scant attention paid to musculoskeletal topics by CoEs may be due to the fact that these are seen as specialty care issues rather than primary care issues. In addition, until recently, medical education has paid little attention to musculoskeletal topics. The definition of women's health care must broaden to include the topics (especially in musculoskeletal health) that impact the quality of life for women.

Given this track record in women's health, it is not surprising that although the evolution of men's health centers started a bit later, it has followed a trajectory similar to that of women's health centers. For men, the reproductive focus was initially on erectile dysfunction; it has now evolved in many centers to include prostate care [15]. Although other centers have focused on cardiac care and problems that threaten life expectancy, men, like women, have not received assistance in disability prevention.

Summary

Sex-based centers of care have played a critical role in improving the health of their constituents and simultaneously drawing attention to the different health care needs of men and women. It is time, however, to challenge them to do an even better job. As other articles in this journal have underscored, there are now overwhelming data that men and women are profoundly different at the molecular and cellular level in virtually all aspects of musculoskeletal health and disease, but the clinical implications of these differences have generally been unexplored. Sex-based health centers of care can play a critical role in exploring these differences and, in doing so, reduce disability and enhance quality of life in our growing population of senior men and women.

References

[1] Collins KS. Evaluation of the National Centers of Excellence in Women's Health: sustaining the promise. Womens Health Issues 2002;12:287–90.

[2] Weitz TA, Freund KM, Wright L. Identifying and caring for underserved populations: experience of the National Centers of Excellence in Women's Health. J Womens Health Gend Based Med 2001; 10:937–52.

[3] Henderson JT, Scholle SH, Weisman CS, et al. The role of physician gender in the evaluation of the National Centers of Excellence in Women's Health: test of an alternate hypothesis. Womens Health Issues 2004;14:130–9.

[4] Harpole LH, Mort EA, Freund KM, et al. A comparison of the preventive health care provided by women's health centers and general internal medicine practices. J Gen Intern Med 2000;15:1–7.

[5] Phelan EA, Burke W, Deyo RA, et al. Delivery of primary care to women: do women's health centers do it better? J Gen Intern Med 2000;15:8–15.

[6] Flocke SA, Gilchrist V. Physician and patient gender concordance and the delivery of comprehensive clinical preventive services. Med Care 2005;43:486–92.

[7] Lurie N, Slater J, McGovern P, et al. Preventive care for women-does the sex of the physician matter? N Engl J Med 1993;329:478–82.

[8] Anderson RT, Weisman CS, Scholle SH, et al. Evaluation of the quality of care in the clinical care centers of the National Centers of Excellence in Women's Health. Womens Health Issues 2002;12: 309–26.

[9] Henrich JB, Chambers JT, Steiner JL. Development of an interdisciplinary women's health training model. Acad Med 2003;78:877–84.

[10] Fife RS, Moskovic C, Dynak H, et al. Development and implementation of novel community outreach methods in women's health issues: the National Centers of Excellence in Women's Health. J Womens Health Gend Based Med 2001;10:27–37.

[11] Tannenbaum C, Mayo N. Women's health priorities and perceptions of care: a survey to identify opportunities for improving preventative health care delivery for older women. Age Ageing 2003;32:626–35.

[12] Mosca L, Allen C, Fernandez-Repollet E, et al. Setting a local research agenda for women's health: the National Centers of Excellence in Women's Health. J Womens Health Gend Based Med 2001; 10:927–35.

[13] Women's Health Care Competencies for Medical Students. Taking steps to include sex and gender differences in the curriculum. Available at: www.apgo.org/wheocomp.

[14] Association of American Medical Colleges. AAMC memorandum 05–12. March 18, 2005.

[15] Association of American Medical Colleges. Medical school objectives project. Report VII: contemporary issues in medicine: musculoskeletal medicine education. 2005.

Index

Note: Page numbers of article titles are in **boldface** type.

A

AAOS. See *American Academy of Orthopaedic Surgeons (AAOS)*.

ACL. See *Anterior cruciate ligament (ACL)*.

Adhesive capsulitis. See *Frozen shoulder*.

Adolescent idiopathic scoliosis (AIS)
    body habitus and, 556
    clinical presentation of, 555–556
    curve pattern in, 556
    curve progression in, 556
    curve stiffness in, 556
    diagnosis of, 555–556
    epidemiology of, 555
    genetic influences on, 555
    growth and development effects of, 555
    natural history of, 556
    sexual dimorphism in, **555–558**
    treatment of, 556–557

Age
    as factor in arthritis, 518
    as factor in musculoskeletal system disabilities, 518–519
    as factor in osteoarthritis of hand, 545
    as factor in osteoporosis, 518–519

AIS. See *Adolescent idiopathic scoliosis (AIS)*.

Amenorrhea, 575

American Academy of Orthopaedic Surgeons (AAOS), 523

Analgesic(s), for musculoskeletal pain, 526

Ankle(s), sexual dimorphism of, **569–574**

Anterior cruciate ligament (ACL)
    biology of, **585–591**
    described, 585
    hormones and, 588–589
    injuries of, factors in, relationships among, 586–587
    sex and, 588–589
    tissue remodeling for, effects of, 587–588

Arthritis
    age-related, 518
    patellofemoral, 596–597

Autoimmune diseases
    sex differences in, **629–633**
        explanations for, 631
        pathogenesis of, 629–631
        sex ratios in, 630

B

Back pain, low, during pregnancy, 551

BMD. See *Bone mineral density (BMD)*.

Body weight, as factor in osteoporosis, 563–564

Bone mineral density (BMD)
    hip fracture and, 615–616
    hormonal influence on, 576
    in osteoporosis, 601

Bone strength
    fractures and, 524–525
    hip fracture and, 615–616

C

Cancer, sexual dimorphism in, 526

Capsulitis, adhesive. See *Frozen shoulder*.

Cartilage, of foot, 571

D

Degenerative disorders of spine, sexual dimorphism in, **549–553**. See also *Sexual dimorphism, in degenerative disorders of spine*.

Developmental dysplasia of hip, 562–563

Dimorphism, **593–599**
    sexual. See *Sexual dimorphism*.

Disability(ies)
    in older adults, 517–519
    of musculoskeletal system, age-related, 518–519

Dislocation(s), patellofemoral, 594–596

E

Eating disorders, in females, 575

Elderly, disabilities in, 517–519

Estrogen, postmenopausal, osteoporosis and, 563

Exercise
 in musculoskeletal disease prevention, 519–520
 osteoporosis effects of, 563

F

Fat mass, hip fracture and, 616

Female athlete triad
 components of, 575–576
 defined, 575–576
 stress fractures and, **575–583**

Foot (feet)
 cartilage of, 571
 extrinsic factors, 571–572
 gait and, 571
 injuries of, shoe effects on, 573
 length of, as proportion of stature for men and women, 570
 ligamentous laxity of, 571
 muscle of, 571
 osteology of, 569–571
 sexual dimorphism of, **569–574**
 shape of, 569–571

Fracture(s)
 bone strength and, 524–525
 hip, **611–622**. See also *Hip(s), fractures of.*
 stress. See *Stress fractures.*

Frozen shoulder, **531–539**
 anatomy of, 531–533
 diagnosis of, 533
 epidemiology of, 531
 physiology of, 531–533
 treatment of
  nonoperative, 533–536
  operative, 536–537

G

Gait, foot in, 571

Gender
 as factor in health research, **513–521**
 as factor in musculoskeletal health, workshop report of, **523–529**
 defined, 593

Genetics, as factor in osteoarthritis of hand, 543

Growth and development
 AIS effects on, 555
 of musculoskeletal system, 524

H

Hand(s), osteoarthritis of, sexual dimorphism of, **541–548**. See also *Osteoarthritis, of hand.*

Health care, quality of, challenges facing, 516–520

Health research
 longevity and, 516–520
 musculoskeletal, future directions in, 515–516
 sex and gender in, **513–521**

Heredity, as factor in osteoarthritis of hand, 543–544

Hip(s)
 developmental dysplasia of, 562–563
 fractures of, **611–622**
  BMD and, 615–616
  bone strength and, 615–616
  described, 611–612
  factors associated with, 614
  management of, 616–617
   pharmacologic, 616–617
   rehabilitation in, 616
  morbidity associated with, 612–614
  mortality associated with, 612
  muscle and fat mass and, 616
  outcomes of, 612–616
  physical function effects of, 614–615
  prevalence of, 611
  psychosocial factors associated with, 615
  sexual predilection for, 611
 injuries of, shoe effects on, 573
 osteoporosis of, sex and gender differences in, **559–568**. See also *Osteoporosis, of hip and knee, sex and gender dierences in.*

Hormone(s)
 ACL and, 588–589
 as factor in osteoarthritis of hand, 544
 BMD effects of, 576

I

Institute of Medicine (IOM), 523

IOM. See *Institute of Medicine (IOM).*

K

Knee(s)
 injuries of
  prior, osteoporosis of hip and knee due to, 564

shoe effects on, 573
osteoporosis of, sex and gender differences in, **559–568**. See also *Osteoporosis, of hip and knee, sex and gender dierences in.*

L

Life expectancy, increased, 516–517

Ligament(s), injuries of, 525

Low back pain, during pregnancy, 551

M

Muscle(s)
  injuries of, 525
  of foot, 571

Muscle mass, hip fracture and, 616

Musculoskeletal diseases, prevention of, physical activity in, 519–520

Musculoskeletal health. See also *Musculoskeletal system.*
  agenda for, 526–527
  gender effects on, workshop report of, **523–529**

Musculoskeletal research, future directions in, 515–516

Musculoskeletal system
  cellular biology of, 524
  disabilities of, age-related, 518–519
  growth and development of, 524
  matrix biology of, 524
  molecular biology of, 524
  pain of, analgesics for, 526

N

National Institutes of Arthritis and Musculoskeletal and Skin Diseases, of NIH, 523

National Institutes of Health (NIH), 564
  National Institutes of Arthritis and Musculoskeletal and Skin Diseases of, 523
  Office of Research in Women's Health of, 523

National Osteoporosis Foundation, 611, 616

Neurologic impairment, after stroke, 626–627

Neurovascular disease, sexual dimorphism in, 525–526

NIH. See *National Institutes of Health (NIH).*

Nutrition, as factor in osteoarthritis of hand, 546

O

Office of Research in Women's Health, of NIH, 523

Osteoarthritis
  genetics of, sexual dimorphism in, 543–544
  of hand
    age as factor in, 545
    anatomy and, 542–543
    concurrent disease and, 545
    epidemiology of, 541–542
    heredity and, 543–544
    hormone status and, 544
    nutrition and, 546
    osteoporosis and, 544–545
    physiology and, 542–543
    risk factors for, 543–546
    sexual dimorphism of, **541–548**
    weight as factor in, 545–546
  osteoporosis and, 561–562

Osteoporosis
  age-related, 518–519
  anatomy associated with, 602–603
  as factor in osteoarthritis of hand, 544–545
  BMD and, 601
  defined, 601
  described, 601
  diagnosis of, 604–605
  epidemiology of, 601–602
  in females, 575–576
    vs. males, **601–609**
  of hip and knee, sex and gender differences in, **559–568**
    anatomy and physiology of, 559–560
    body weight and, 563–564
    developmental dysplasia of hip and, 562–563
    diagnosis of, 564–565
    epidemiology of, 559
    exercise and, 563
    in prevention, 561–564
    postmenopausal estrogen use, 563
    prior knee injury and, 564
    risk factors for, 561–564
    treatment methods, 565
  osteoarthritis and, 561–562
  pathophysiology of, 560–561
  prevalence of, 601–602
  prevention of, 603–604
  risk factors for, 603–604
  treatment of, 605–607

P

Pain
　back, low, during pregnancy, 551
　musculoskeletal, analgesics for, 526
　patellofemoral, 593–594

Patellofemoral disorders, **593–599**
　arthritis, 596–597
　dislocations, 594–596
　pain associated with, 593–594

Physical activity, in musculoskeletal disease prevention, 519–520

Pregnancy, low back pain during, 551

Q

Quality of health care, challenges facing, 516–520

Quality of life, challenges facing, 516–520

R

Rehabilitation, in hip fracture management, 616

S

Scoliosis, adolescent idiopathic. See *Adolescent idiopathic scoliosis (AIS)*.

Sex
　ACL and, 588–589
　as factor in health research, **513–521**
　defined, 593

Sex differences, in autoimmune diseases, **629–633**
　pathogenesis of, 629–631

Sex-based centers of care, future of, **635–637**

Sexual dimorphism
　in cancer, 526
　in degenerative disorders of spine, **549–553**
　　female predominance, 551–552
　　male predominance, 551
　　perioperative considerations, 552
　in neurovascular disease, 525–526
　in stroke, **623–628**
　of foot and ankle, **569–574**
　of osteoarthritis of hand, **541–548**

Shoe(s), injury effects of, 573

Shoulder(s), frozen, **531–539**. See also *Frozen shoulder*.

Spine
　anatomy of, 549–550
　degenerative disorders of, sexual dimorphism in, **549–553**. See also *Sexual dimorphism, in degenerative disorders of spine*.
　density of, 550–551
　development of, 549–550
　　qualitative, 550–551
　embryology of, 549–550
　maturation of, 549–550

Strengthening activities, in musculoskeletal disease prevention, 520

Stress fractures
　complications of, 580–581
　defined, 576
　described, 576–578
　diagnosis of, 578
　female athlete triad and, **575–583**. See also *Female athlete triad*.
　imaging of, 578
　prevention of, 581
　return to play after, 581
　treatment of, 579–580

Stroke
　acute, treatment of, 624–625
　burden of care after, 626
　chronic mobility and functional impairment after, treatment of, 627
　diagnostic evaluation for, 624
　life expectancy after, 623
　neurologic impairment after, 626–627
　neurologic recovery after, 627
　prevalence of, 623
　preventive treatment for, 624
　recurrent, prevention of, 625–626
　sexual dimorphism in, **623–628**
　susceptibility to, 623–624

T

Tendon(s), injuries of, 525

U

Upper extremity, **531–539**. See also specific disorders of, e.g., *Frozen shoulder*.

W

Weight
　as factor in osteoarthritis of hand, 545–546
　as factor in osteoporosis, 563–564

Women's health research
　current status of, 513–515
　evolution of, 513–515

Moving?

Make sure your subscription moves with you!

To notify us of your new address, find your **Clinics Account Number** (located on your mailing label above your name), and contact customer service at:

E-mail: elspcs@elsevier.com

800-654-2452 (subscribers in the U.S. & Canada)
407-345-4000 (subscribers outside of the U.S. & Canada)

Fax number: 407-363-9661

Elsevier Periodicals Customer Service
6277 Sea Harbor Drive
Orlando, FL 32887-4800

*To ensure uninterrupted delivery of your subscription, please notify us at least 4 weeks in advance of move.

United States Postal Service
Statement of Ownership, Management, and Circulation

1. Publication Title	2. Publication Number	3. Filing Date
Orthopedic Clinics of North America	9 5 0 - 9 2 0	9/15/06

4. Issue Frequency	5. Number of Issues Published Annually	6. Annual Subscription Price
Jan, Apr, Jul, Oct	4	$190.00

7. Complete Mailing Address of Known Office of Publication *(Not printer) (Street, city, county, state, and ZIP+4)*	Contact Person
Elsevier, Inc. 360 Park Avenue South New York, NY 10010-1710	Sarah Carmichael
	Telephone
	(215) 239-3681

8. Complete Mailing Address of Headquarters or General Business Office of Publisher *(Not printer)*

Elsevier, Inc., 360 Park Avenue South, New York, NY 10010-1710

9. Full Names and Complete Mailing Addresses of Publisher, Editor, and Managing Editor *(Do not leave blank)*

Publisher *(Name and complete mailing address)*

John Schrefer, Elsevier, Inc., 1600 John F. Kennedy Blvd., Suite 1800, Philadelphia, PA 19103-2899

Editor *(Name and complete mailing address)*

Debora Dellapena, Elsevier, Inc., 1600 John F. Kennedy Blvd., Suite 1800, Philadelphia, PA 19103-2899

Managing Editor *(Name and complete mailing address)*

Catherine Bewick, Elsevier, Inc., 1600 John F. Kennedy Blvd., Suite 1800, Philadelphia, PA 19103-2899

10. Owner *(Do not leave blank. If the publication is owned by a corporation, give the name and address of the corporation immediately followed by the names and addresses of all stockholders owning or holding 1 percent or more of the total amount of stock. If not owned by a corporation, give the names and addresses of the individual owners. If owned by a partnership or other unincorporated firm, give its name and address as well as those of each individual owner. If the publication is published by a nonprofit organization, give its name and address.)*

Full Name	Complete Mailing Address
Wholly owned subsidiary of	4520 East-West Highway
Reed/Elsevier, US holdings	Bethesda, MD 20814

11. Known Bondholders, Mortgagees, and Other Security Holders Owning or Holding 1 Percent or More of Total Amount of Bonds, Mortgages, or Other Securities. If none, check box → ☐ None

Full Name	Complete Mailing Address
N/A	

12. Tax Status *(For completion by nonprofit organizations authorized to mail at nonprofit rates) (Check one)*
The purpose, function, and nonprofit status of this organization and the exempt status for federal income tax purposes:
☐ Has Not Changed During Preceding 12 Months
☐ Has Changed During Preceding 12 Months *(Publisher must submit explanation of change with this statement)*

(See Instructions on Reverse)

PS Form **3526**, October 1999

13. Publication Title	14. Issue Date for Circulation Data Below
Orthopedic Clinics of North America	July 2006

15.	Extent and Nature of Circulation		Average No. Copies Each Issue During Preceding 12 Months	No. Copies of Single Issue Published Nearest to Filing Date
a.	Total Number of Copies *(Net press run)*		3,800	3,600
b. Paid and/or Requested Circulation	(1)	Paid/Requested Outside-County Mail Subscriptions Stated on Form 3541. *(Include advertiser's proof and exchange copies)*	1,640	1,487
	(2)	Paid In-County Subscriptions Stated on Form 3541 *(Include advertiser's proof and exchange copies)*		
	(3)	Sales Through Dealers and Carriers, Street Vendors, Counter Sales, and Other Non-USPS Paid Distribution	1,110	1,190
	(4)	Other Classes Mailed Through the USPS		
c.	Total Paid and/or Requested Circulation *[Sum of 15b. (1), (2), (3), and (4)]*	▶	2,740	2,677
d. Free Distribution by Mail *(Samples, complimentary, and other free)*	(1)	Outside-County as Stated on Form 3541	140	127
	(2)	In-County as Stated on Form 3541		
	(3)	Other Classes Mailed Through the USPS		
e.	Free Distribution Outside the Mail *(Carriers or other means)*			
f.	Total Free Distribution *(Sum of 15d. and 15e.)*	▶	140	127
g.	Total Distribution *(Sum of 15c. and 15f.)*	▶	2,880	2,804
h.	Copies not Distributed		920	796
i.	Total *(Sum of 15g. and h.)*	▶	3,800	3,600
j.	Percent Paid and/or Requested Circulation *(15c. divided by 15g. times 100)*		95.14%	95.47%

16. Publication of Statement of Ownership
☐ Publication required. Will be printed in the **October 2006** issue of this publication. ☐ Publication not required

17. Signature and Title of Editor, Publisher, Business Manager, or Owner | Date

(signature) John Fanucci — Executive Director of Subscription Services | 9/15/06

I certify that all information furnished on this form is true and complete. I understand that anyone who furnishes false or misleading information on this form or who omits material or information requested on the form may be subject to criminal sanctions (including fines and imprisonment) and/or civil sanctions (including civil penalties).

Instructions to Publishers

1. Complete and file one copy of this form with your postmaster annually on or before October 1. Keep a copy of the completed form for your records.
2. In cases where the stockholder or security holder is a trustee, include in items 10 and 11 the name of the person or corporation for whom the trustee is acting. Also include the names and addresses of individuals who are stockholders who own or hold 1 percent or more of the total amount of bonds, mortgages, or other securities of the publishing corporation. In item 11, if none, check the box. Use blank sheets if more space is required.
3. Be sure to furnish all circulation information called for in item 15. Free circulation must be shown in items 15d, e, and f.
4. Item 15h., Copies not Distributed, must include (1) newsstand copies originally stated on Form 3541, and returned to the publisher, (2) estimated returns from news agents, and (3), copies for office use, leftovers, spoiled, and all other copies not distributed.
5. If the publication had Periodicals authorization as a general or requester publication, this Statement of Ownership, Management, and Circulation must be published; it must be printed in any issue in October or, if the publication is not published during October, the first issue printed after October.
6. In item 16, indicate the date of the issue in which this Statement of Ownership will be published.
7. Item 17 must be signed.

Failure to file or publish a statement of ownership may lead to suspension of Periodicals authorization.

PS Form **3526**, October 1999 *(Reverse)*